S0-BRD-435

SCIENCE AND **TECHNOLOGY** AND THE **FUTURE** **DEVELOPMENT** OF **SOCIETIES**

International Workshop Proceedings

Glenn Schweitzer, *Editor*

Committee on the U.S.-Iran Workshop on Science and Technology and the Future Development of Societies

Office for Central Europe and Eurasia
Development, Security, and Cooperation
Policy and Global Affairs

NATIONAL RESEARCH COUNCIL
OF THE NATIONAL ACADEMIES

THE NATIONAL ACADEMIES PRESS
Washington, D.C.
www.nap.edu

THE NATIONAL ACADEMIES PRESS   500 Fifth Street, N.W.   Washington, DC 20001

NOTICE: The project that is the subject of this report was approved by the Governing Board of the National Research Council, whose members are drawn from the councils of the National Academy of Sciences, the National Academy of Engineering, and the Institute of Medicine. The members of the committee responsible for the report were chosen for their special competences and with regard for appropriate balance.

Any opinions, findings, conclusions, or recommendations expressed in this publication are those of the author(s) and do not necessarily reflect the views of the organizations or agencies that provided support for the project.

International Standard Book Number-13:   978-0-309-12025-8
International Standard Book Number-10:   0-309-12025-X

A limited number of copies are available from the Office for Central Europe and Eurasia, National Research Council, 500 Fifth Street, N.W., Washington, DC 20001; (202) 334-2376.

Additional copies of this report are available from the National Academies Press, 500 Fifth Street, N.W., Lockbox 285, Washington, DC 20055; (800) 624-6242 or (202) 334-3313 (in the Washington metropolitan area); Internet, http://www.nap.edu.

Printed in the United States of America.

THE NATIONAL ACADEMIES
Advisers to the Nation on Science, Engineering, and Medicine

The **National Academy of Sciences** is a private, nonprofit, self-perpetuating society of distinguished scholars engaged in scientific and engineering research, dedicated to the furtherance of science and technology and to their use for the general welfare. Upon the authority of the charter granted to it by the Congress in 1863, the Academy has a mandate that requires it to advise the federal government on scientific and technical matters. Dr. Ralph J. Cicerone is president of the National Academy of Sciences.

The **National Academy of Engineering** was established in 1964, under the charter of the National Academy of Sciences, as a parallel organization of outstanding engineers. It is autonomous in its administration and in the selection of its members, sharing with the National Academy of Sciences the responsibility for advising the federal government. The National Academy of Engineering also sponsors engineering programs aimed at meeting national needs, encourages education and research, and recognizes the superior achievements of engineers. Dr. Charles M. Vest is president of the National Academy of Engineering.

The **Institute of Medicine** was established in 1970 by the National Academy of Sciences to secure the services of eminent members of appropriate professions in the examination of policy matters pertaining to the health of the public. The Institute acts under the responsibility given to the National Academy of Sciences by its congressional charter to be an adviser to the federal government and, upon its own initiative, to identify issues of medical care, research, and education. Dr. Harvey V. Fineberg is president of the Institute of Medicine.

The **National Research Council** was organized by the National Academy of Sciences in 1916 to associate the broad community of science and technology with the Academy's purposes of furthering knowledge and advising the federal government. Functioning in accordance with general policies determined by the Academy, the Council has become the principal operating agency of both the National Academy of Sciences and the National Academy of Engineering in providing services to the government, the public, and the scientific and engineering communities. The Council is administered jointly by both Academies and the Institute of Medicine. Dr. Ralph J. Cicerone and Dr. Charles M. Vest are chair and vice chair, respectively, of the National Research Council.

www.national-academies.org

Preface

In June 2006, seventeen scientists and educators selected by the National Academies, the Academy of Sciences of Iran, and the Académie des Sciences of France held a workshop at the estate of the Fondation des Treilles in Toutour, France, to discuss issues concerning the role of science in the development of modern societies. This location was an idyllic setting for relaxed conversations, while simplifying travel and visa arrangements for the participants. The three academies had organized a related workshop on food safety and security, global energy transitions, and education issues at the estate in 2003; the success of that workshop was a strong incentive to return to Toutour.

This report includes the presentations made at the workshop and summarizes the discussions that followed the presentations. The statements made in the enclosed papers are those of the individual authors and do not necessarily represent positions of the National Academies. Unfortunately, three of the Iranian specialists and one American specialist who were scheduled to participate were unable to travel to France. However, they provided the papers that they intended to present. These papers are included in the appendixes of the report.

An important observation of the participants was that the topics that were addressed, particularly the inclusion of science-oriented themes throughout the K-12 education curriculum, warranted more detailed discussions between Iranian and Western colleagues. The National Academies are currently exploring opportunities to continue such discussions in Iran, and this report provides useful background for the further development of interactions of Western scientists and educators with Iranian specialists.

ACKNOWLEDGMENTS

This volume has been reviewed in draft form by individuals chosen for their technical expertise, in accordance with procedures approved by the National Research Council's Report Review Committee. The purpose of such an independent review is to provide candid and critical comments that will assist the institution in making its published report as sound as possible and to ensure that the report meets institutional standards for quality. The review comments and draft manuscript remain confidential to protect the integrity of the process.

We wish to thank the following individuals for their review of selected papers: R. Stephen Berry, University of Chicago; James Childress, University of Virginia; Denis Gray, North Carolina State University; Janet Hustler, Synopsys, Inc.; Richard McCray, University of Colorado at Boulder; Wilhelmine Miller, George Washington University; and Andrew Schrank, University of New Mexico.

Although the reviewers listed above have provided constructive comments and suggestions, they were not asked to endorse the content of the individual papers. Responsibility for the final content of the papers rests with the individual authors and the institution.

George Bugliarello
Chair, National Research Council Committee on
the U.S.-Iran Workshop on Science and Technology
and the Future Development of Societies

Glenn E. Schweitzer
Director, Office for Central Europe and Eurasia,
National Research Council

Contents

APPENDIXES

SCIENCE AND SOCIETY ISSUES

The Role of Communications and Scientific Thinking

BARBARA SCHAAL
Washington University

The international scientific community faces many challenges, from the funding of scientific research and the recruiting of new individuals into the fields of science and technology to the communication of both the value of science and specific scientific information. Communication of science is a topic that is frequently omitted in the discussion of science and science policy, yet scientific information is often ignored in policy decisions as a result of a failure in communication. This failure can have profound consequences. It is very clear that the future economic development of many countries increasingly rests on innovations in science and technology. Without adequate scientific input into policy decisions, future development may be hindered. A challenge for both national scientific groups and the international scientific community is to communicate the importance of science and its direct role in economic and social development. An even greater challenge is how to effectively communicate scientific information to decision makers in such a way that policy decisions are based on sound science.

Scientific groups, such as national academies of sciences, professional societies, and research institutions and universities, all have particular, varied strengths and limited financial and personnel resources. Here we consider how national academies can effectively communicate the importance of science and scientific thinking to decision makers who are not scientifically trained. In addition, we address how scientists can effectively communicate to decision makers the specific scientific information that is essential for sound science policy. Effective communication programs for any organization go through a prescribed planning process that includes identifying goals and tactics. Planning includes addressing such questions as: What do we want to communicate? To whom do we want to communicate? How do we communicate information? Such an exercise is extremely useful in identifying goals, setting priorities for communication, and effectively using limited resources.

Several recent studies and popular books have made the claim that the globe is becoming increasingly flat (Friedman, 2005). Past reports from the U.S. National Research Council have stated that decision making in our increasingly complex and connected world environment will be based on complex, inter-disciplinary scientific research and that decision makers and stakeholders will need to have increased involvement with science in order to make appropriate policy decisions.

For interactions between scientific organizations and decision makers to be successful, several components are necessary. Jacobs identifies areas that need to be considered for effective communication of scientific information (Jacobs, 2003). The first is understanding what information is needed for a policy decision and understanding the perspective and context of the "client." Just as in any communication initiative, it is essential to understand who the audience is. Second is the need to understand the mechanisms of communication for an effective collaboration between scientific organizations and decision makers. Science groups should develop a communication strategy to ensure open and effective dialog. Issues such as the usability of information and equity of benefits need to be discussed. Third, incentives for change need to be considered. In many past cases, interactions between scientists and decision makers have been ineffective. What are the incentives to work together? Are there common goals? Fourth, it is essential that mechanisms for evaluation, feedback, and measures of success be put in place. Without a dialog between scientists and policy makers, the quality and usefulness of scientific information will not be improved. Is the scientific information adequate and has it been provided in a form that is useful?

National academies of sciences have a unique role to play in national and international science policy. National academies have the prestige to speak for a country's scientific establishment, and they are often the source of the most current scientific information, information that is increasingly important for policy decisions. Communicating that information in an effective and useful manner to decision makers will increase the quality and usefulness of the policy decisions that are essential for a nation's future development.

REFERENCES

Friedman, T. 2005. *The World Is Flat: A Brief History of the Twenty-First Century.* New York: Farrar, Straus, and Giroux.

Jacobs, K. 2003. *Connecting Science, Policy, and Decision-Making: A Handbook for Researchers and Science Agencies.* Silver Spring, MD: National Oceanic and Atmospheric Administration Office of Global Programs.

Knowledge, Validation, and Transfer: Science, Communication, and Economic Development

JOHN ENDERBY
Institute of Physics

Science, engineering, technology, and innovation (SET&I) are the bedrock of a successful economy, particularly as nations are moving toward knowledge-based economies. Communication of SET&I knowledge plays a fundamental role in shaping policy on science-related issues and can be considered a driving force for socioeconomic development. However, effective science communication is not simple. The United Kingdom has experienced its share of challenges in dealing with controversial issues such as genetically modified foods or bovine spongiform encephalopathy and has learned many lessons in attempting to engage the general public with science. Science and technology continue to advance at an increasingly rapid rate, and discussion of the issues that arise from these developments is highly important.

SET&I are particularly important for developing countries in order to raise living standards, create wealth, and ensure that their natural resources and biodiversity are not degraded. They also underpin the majority of the Millennium Development Goals. Developing countries face numerous challenges to development. There has been extensive land degradation through deforestation and overcultivation. There exists a scarcity of safe drinking water in numerous countries, and high rates of disease have a profound effect on their economies. Climate change could be a serious obstacle for economic and human development and has the potential to adversely affect gains already made. SET&I have a role to play in addressing these challenges. Communicating the ways in which SET&I will address these issues to all levels of a nation's society is crucial. We therefore begin by considering access to validated knowledge generated by primary research papers.

FULL AND FAIR ACCESS TO SCIENTIFIC LITERATURE

In developed countries it can be assumed that investigators will have access to the primary literature as both readers and authors. Moreover, such literature will have undergone quality control through the mechanism of peer review. It is becoming increasingly recognized that scientists in developing countries will be placed at a major disadvantage if similar rights do not exist for them. First, in a subject that is rapidly developing, time, money, and effort could be wasted if only yesterday's science is available. Second, high-quality work carried out by authors in developing countries should, if the authors so wish, be published in highly cited and read international journals. This two-way exchange of information can only assist the community of scientists worldwide.

There have been many business models proposed to accomplish these most laudable aims, and these have been reviewed in the excellent book by John Willinsky (2005). At one extreme is the *author pays/open access* model pioneered by the Institute of Physics *New Journal of Physics* and the U.S.-based Public Library of Science, among others.

The advantages of this model are obvious so far as readers are concerned; it is less obvious whether this model is advantageous to authors from developing countries or to authors who have not received research funding. One difficulty is that high-prestige/high-rejection-rate journals need to cover their costs by levying substantial charges on successful authors. Some have suggested that authors should pay part of the fee on submission and the rest on acceptance, but this may lead to increased bureaucracy at the institutional level.

At the other extreme is the *subscription model* in which the reader (or more often the parent institution) pays for access. This model gives advantage to authors because their work can be submitted and assessed free of charge. On the other hand the inexorable rise in subscription rates leads to cancellations and loss of access. The International Network for the Availability of Scientific Publications and the Health InterNetwork Access to Research Initiative (HINARI) are addressing this issue for developing countries with considerable success; an example of the work of HINARI in Kenya can be found in Willinsky's (2005) *Access Principle.*

Between these two extremes are a variety of hybrid models. It is important, in my personal view, not to become too obsessed with one particular model. Rather, one should explore all possibilities, not least those opened up by the World Wide Web, to ensure the fundamental right for full and fair access, which I define as the opportunity to *read* research papers and to *submit* research papers decoupled from the level of economic development.

ACCESS TO OVERLOOKED MATERIAL

In the debate about open access, patent and gray literature are often overlooked. The free-access Web site http://www.scidev.net provides news, views, and information on issues relating to science, engineering, and technology (SET). As far as innovation is concerned, patent literature is of particular importance. Patents are of course just one example of materials subject to intellectual property (IP) protection, which includes copyright, trademarks, industrial designs, breeders' rights, and so forth.

Again, the situation is developing and the Royal Society (2003) report, *Keeping Science Open,* made several recommendations of relevance to developing countries. For example, it recommended that developing countries should not be required to implement trade-related aspects of intellectual property rights until their state of development is such that the stimulation effect on innovation will be worth the costs and restraints inherent in IP systems. It further recommended that the World Intellectual Property Organization should continue to work with governments to provide guidelines for informed consent and profit sharing that can be translated into practical situations involving the exploitation of traditional knowledge for the benefits of the holders of such knowledge and humankind in general.

THE SUPPRESSION OF RESEARCH RESULTS

As a general principle the results of research should be in the public domain, even if negative or null conclusions follow (e.g., in drug trials). There may be occasions, however, when the results of research need to be withheld, perhaps for a finite period. Issues relating to national defense and security, law enforcement, or trade secrets could, on limited occasions, be the subject of restrictions; but the presumption must be that scientific research must be open for scrutiny unless cogent and compelling reasons dictate otherwise.

SCIENCE AND THE PUBLIC INTEREST

Knowledge sharing must increase people's awareness of information developed through science that can help them in their daily lives. It must also be multidirectional. Engagement in SET&I issues by nonscientists will help to stimulate open dialogue, mutual understanding, and consensus between the scientific and nonscientific communities. Technologies, if harnessed appropriately, could have the potential for countries to make substantial strides in development. Emerging sciences such as nanotechnology, biotechnology, new materials, and information communication technology will have profound implications for long-term economic growth. It will be important for governments to engage all members of society in the issues surrounding the application of these technologies. Effective science education also has a role in strengthening the science base. All stakeholders in the SET enterprise must be sensitive to the concerns of fellow scientists and those of the general public. The Royal Society (2006) has produced a report on good practice in communicating new scientific research to the public. It proposed a checklist of some 15 items that could help researchers take into account public interest and to avoid some of the difficulties scientists have had in the past with science deemed controversial by a variety of interested parties.

SET AND ITS APPEAL TO YOUNGER PEOPLE

If SET are to prosper, and if it is believed, as most countries believe, that they are an important component of economic development, then a good supply of talented individuals going into science is a necessity. In this regard, the Relevance of Science Education (ROSE) study is of particular interest. Schreiner and Sjøberg (2004) report fully on the project rationale, development, and logistics. A useful summary can be found in an article by Sjøberg and Schreiner (2006). There are dramatic differences between students in developed and in developing countries. In developing countries, students have a strong desire to take part in SET. In most developed countries, students are, on average, much less enthusiastic. Moreover, the gender bias against SET is far more marked in the northern countries of the Organisation for Economic Co-operation and Development than elsewhere. Clearly, this study raises concerns for Europe and the United States, but at the same time is encouraging for countries undergoing economic development. There are policy implications for both developed and developing counties. In the spirit of this paper, the rich countries of the West have a duty to ensure that scientists from developing countries are not disadvantaged, either by design or by accident, from actions such as unnecessary restrictions on

- travel,
- the free and full access to validated knowledge, and
- the ability to set up appropriate collaborations.

Let me finally quote Paul Schlumberger in writing to his son, Conrad, in 1915. He summarizes how the SET community should approach some of the issues raised at this meeting:

If the convergence of scientific and commercial viewpoints is too difficult, it is better to opt for the viewpoint of Science. Science is a great force for peace, for the individual as well as for humanity.

REFERENCES

Royal Society. 2003. Keeping science open: The effects of intellectual property policy on the conduct of science. Available at: http://royalsociety.org/displaypagedoc.asp?id=11403. Accessed March 30, 2008.

Royal Society. 2006. Science and the public interest: Communicating the results of new scientific research to the public. Available at: http://royalsociety.org/page.asp?tip=1&id=6982. Accessed March 30, 2008.

Schreiner, C., and S. Sjøberg. 2004. Sowing the seeds of ROSE: Background, rationale, questionnaire development, and data collection for ROSE—A comparative study of students' views of science and science education. ROSE: The relevance of science education. Available at: http://www.ils. uio.no/forskning/publikasjoner/actadidactica/english.html. Accessed March 30, 2008.

Sjøberg, S., and C. Schreiner. 2006. How do students perceive science and technology? *Science in School* 1(Spring):66–69. Available at: http://www.scienceinschool.org/2006/issue1/rose/. Accessed March 30, 2008.

Willinsky, J. 2005. *The Access Principle: The Case for Open Access to Research and Scholarship.* Cambridge, MA: MIT Press.

The Morality of Exact Sciences

YOUSEF SOBOUTI

Institute for Advanced Studies in Basic Sciences

With due respect to the sayings of the sages of old and modern times, I would like to offer my own definitions of culture and morality. In my view, the mix of the beliefs and deeds of a society constitutes the culture of that society. The consent and consensus among the members of a society to respect and practice certain forms of behavior and conduct form the moral codes of that society.

Environmental factors are among the main, if not the sole, factors shaping cultures. For instance, the Bedouin Arab and the desert-dwelling Iranian of the wind- and sand-stricken drylands have to protect themselves from the scorching sun of their habitat by covering virtually all parts of their bodies. On the other hand, the inhabitants of the wet tropics have to minimize their clothing to enhance the ventilation of the body, and the Europeans of the misty green continent find it a health requirement to expose themselves to the rare and much-sought-after sunshine wherever and whenever they happen to find it. In the course of centuries and millennia, these practices become ingrained and eventually emerge as unyielding traditions and even religious beliefs demanding strict observance. Then there comes a time when a young Muslim girl is required to remove her head scarf in a French school. She feels insulted and her religious convictions violated. Just the same, when a Western woman in Tehran's bazaar is asked to cover herself up properly, she too feels offended and deprived of her basic rights.

With the definition I have given, it is natural to expect that human societies separated from each other, either geographically or in time, have different cultures and different codes of morality. Conflicts and atrocities may arise when different cultures come together.

Until about a century ago, interactions between societies took place mainly through trade and wars. Both of these mechanisms, however, operated on a much smaller scale than they do today. There were travelers, students, Sufis, and other adventurers who moved from one place to another and helped cultural exchanges. These people were few in number and left little impact on the societies they visited. At any rate, cultural acquisitions from others were gradual and slow. Societies had ample time to adapt to whatever changes were deemed necessary or inevitable.

Modern science and its offspring, modern technology, have drastically changed the situation. A journey from Morocco to India or from Europe to China in the thirteenth century, which took Ebne Batootah and Marco Polo years to complete, now is at the reach of hundreds of thousands of air travelers and can be accomplished within hours. A caravan load of merchandise from Khatai to the coasts of Genoa, which had to be sold and resold several times along the Silk Road before reaching its destination, has in our time been replaced by businesses that are capable of moving thousands of tons of goods across the world within days. Most astonishing of all is the speed of the exchange of information. In 400 B.C., Darius the Great took pride in having created a system of communication that could deliver his order from Persepolis to his satrap in Lydia within a day. This, in the early years of the twenty-first century, has been made possible by the instantaneous transmission of information on coded electromagnetic signals in volumes of giga- and terabytes.

The plain fact that interactions between societies take place on a much larger scale and in a much shorter time frame makes societies prone to tension. Cultures don't find enough time to adjust themselves to changes dictated by modern-day science and technology. This is irrespective of whether the changes are desired and sought-after or not desired and resented. I give an example of each case.

No one disputes the values of modern hygiene and medicine. Its widespread use, however, has caused a worldwide population explosion, particularly in developing societies. As a solution to the problem, marrying at a young age is less common, and celibacy up to the ages of 30–35 has become the order of the day. This remedy, in turn, has created a new sort of problem in Muslim societies wherein sexual relations are allowed only through legal and publicly announced marriages.

As a second example, the widespread use of mass media (newspapers, radio, television, telephone, fax, the Internet) helps to inform people. However, informed minds are inquisitive creatures. They poke their noses into whatever they come across. There are numerous societies where the ruling clan, whether elected or inherited, detests interferences.

To summarize, cultures are largely influenced by environmental factors that are mainly non-human. Every culture has its own moral codes. Cultures and morals are dynamic systems and evolve in time. However, like any other dynamic system, they have inertia, and they resist changes. Modern technology and communication systems of our time are the main factors demanding changes and imposing strains on morals.

Is there a way to cope with such strains and to prevent crises within societies and clashes between cultures? In my opinion, there is. In the course of the past two to three centuries, exact sciences have developed a way to reason out differences in opinions and bring about consensus without resorting to atrocities. Perhaps this procedure could be used effectively to settle the disputes that at first glance might look nonscientific.

By definition a science is exact if (a) it draws its principles and axioms from observations of the phenomena occurring in nature, (b) it uses mathematical logic to draw conclusions from its founding principles, and (c) it checks the validity of its conclusions by subjecting them to experiments. In this procedure there is no place for the beliefs and convictions of the scientist. A scientist, no matter how great his or her achievements, is never promoted to a state of scientific sainthood. The scientist and his or her opinions remain subject to criticism. In exact sciences, human interference is minimized. The success lies in identifying the causes of the effects. Once this identification is achieved, it takes little ingenuity to solve the problem at hand.

Let us call this scientific method *rationality* and *rational thinking*. There was a time in the history of mankind when undesirables were attributed to evil forces of nature. In the course of time man distanced himself from this notion but placed the evil in the minds of his fellow humans. The latter conviction persists. In crises that are not scientifically and rationally analyzed, people often find evil doings and inevitably resort to violence to resolve the disputes. For example, physics, chemistry, and to a certain extent biology have long become axiomatic and exact. Professionals in these disciplines don't settle their differences of opinion by accusing each other of heresy, sorcery, and so forth.

Before these disciplines became axiomatic the situation was different. Giordano Bruno was accused of sorcery and was burned. Nicolaus Copernicus could not publish his book, *De Revolutionibus Orbium Coelestium*, on the heliocentric theory of the skies in his lifetime for fear of his colleagues. Galileo, however, was wise enough to deny the motion of the Earth and save his neck. In Eastern intellectual circles, Imam Ghazali[1] pronounced Farabi, Avicenna, and Ibne Rushd (also called Averroes) as heretics because he did not agree with their perception of natural philosophy. Sohravardi was condemned to death by his colleagues, again because of his philosophical point of view.

In our time, the science of economics and the art of managing governments, legislative and judicial systems, as well as issues of human rights and so forth, do not fall in the category of exact and axiomatic disciplines. They do not have a tension-free and rationality-based mechanism to settle disputes. It is my strong

---

[1]Abu Hamid Muhammad ibn Muhammad Al-Ghazali (1058–1111).

conviction that mass dissemination of rational thinking through the promotion of science education in all societies is helpful in reducing global tensions and in opening doors to logical reasoning instead of presenting human beliefs as evidence of rightfulness.

Science and Society Issues: Summary of Discussion

NORMAN NEUREITER

American Association for the Advancement of Science

As it was pointed out in the discussion, in the United States it is considered extremely important to develop strategies to convey the importance of science and technology for the continued growth and prosperity of the United States to policy makers, Congress, and the presidential administration. This is especially true in the face of challenges to the U.S. technological leadership position from growing competition around the world and the lack of interest in science and technology among U.S. students. There are also religious and ideological objections to certain aspects of scientific investigation.

Different approaches for conveying to the U.S. government the importance of science and technology to continued U.S. growth and prosperity were discussed, such as the media, direct impacts with decision makers, and possibly the development and use of the Internet. Also noted were the difficulties in communicating the uncertainties of science, such as incomplete data.

In discussion, it was pointed out that the impact of the community or individual views on Iranian decision makers regarding science and technology is very minimal. However, there is not an anti-science mood in Iran among students. Islam becomes most involved in the creationism-evolution dispute. Also, where societies and people are poor, the sophisticated debates over science (e.g., those involving genetically modified organisms) are not relevant. The issues at hand are food and survival.

Furthermore, although it is ideal for decision makers to decide questions only after rational analysis, there is a tendency to make decisions based on emotion or attitude. This implies that a mentality change is needed among decision makers.

Regarding the validation and transfer of scientific research, research is not complete until it is peer reviewed and published; that is the validation process of science. There should be full and fair access to knowledge for all; the subscription model, where journals go into public libraries, is the right approach.

Concern was expressed about trade-related aspects of rules for intellectual property rights (Agreement on Trade Related Aspects of Intellectual Property Rights, or TRIPS agreement), including copyrights and patents and their impact on developing countries. Examples were provided of exploitation of genetic resources of developing countries, and so the importance of protecting such property was recognized. It was stated that on big issues, such as global warming, we must have international cooperation, for example, on carbon dioxide sequestration, wherein each country would give up its intellectual property rights for the global good.

Science is an activity where the practitioners do not kill each other when they disagree. Habits derived from necessities over time (diet, clothing, and so forth) can become traditions or doctrinal regulations. When crises are not analyzed, people attribute evil to the other side. Scientific education can lead to rational resolutions of conflict; however, change can be traumatic for people, and as such is often opposed.

Government systems are not precise systems, and free elections are not always the optimum answer for a society. The question was raised whether human rights are the same for the United States, China, Sweden, and Saudi Arabia.

In addition, it was noted that habits do not have to become traditions; they can change. Rationality often does not apply in human affairs, despite scientists' pleas for it to be used. One must be sensitive to different cultures and questions regarding the universality of any given system of values. Any major change challenges the culture, habits, or traditions of people and hence is met with reluctance or resistance.

THE ROLE OF
SCIENCE AND ENGINEERING
IN DEVELOPMENT

Women in Academic Science and Engineering in the United States: Challenges and Opportunities

GERALDINE RICHMOND
University of Oregon

As the global economy becomes increasingly more technological, it is important that we recruit and employ science and engineering talent from all sectors of our population, regardless of gender, race, or creed. In many countries, discriminatory practices have limited the participation of many groups in the global science and engineering enterprise. In the United States, women are increasingly majoring in science and engineering fields, and more women are earning graduate degrees in these fields. However, women are underrepresented in a number of important fields. In biological sciences, women earn almost one-half of the undergraduate and graduate degrees, whereas in engineering, women earn less than 20 percent of undergraduate and graduate degrees.

Though progress toward the goal of parity in the workforce has occurred in recent years, it has been too slow. Women scientists and engineers continue to experience greater difficulty building academic careers than men of comparable training and background, with the greatest discrepancies at research-intensive universities and at the higher academic ranks. Bias against women and caregivers has long existed in universities, particularly for women who enter stereotypically male occupations such as science and engineering. In addition, women must contend with practices and policies that appear neutral but that disadvantage women compared with men. Though in many cases the advantages that males receive are small, they nonetheless accumulate over time into large differences in recognition and prestige.

In U.S. universities, women's representation decreases with each step up the tenure-track and academic leadership hierarchy. In most fields, the decrease is out of proportion to Ph.D. degree production even 10 or 15 years ago. Increasing representation is not just a matter of catch-up time. Proportionately fewer women apply for tenure-track positions in science and engineering. In most science and engineering fields, women are under-represented on university faculties with respect to the number of Ph.D.'s produced. Isolation is a major problem for all women in academic ranks, and it results largely from their small numbers, particularly for women of color. In short, women scientists and engineers in the United States from minority racial or ethnic backgrounds must contend with obstacles even more severe than those of their white colleagues.

While there is increasing representation of minority women in undergraduate and graduate programs, their numbers do not come close to their representation in the general population. In 2004, African Americans earned only 2.5 percent of all doctorates in the biological sciences and only 4.5 percent of engineering doctorates. African American women earned the majority of these doctorates, and yet, these women are less represented in academic faculties than are African American men. Minority women are almost completely unrepresented on faculties of science and engineering at research universities. In 2002, there were a total of 94 African American, 53 Hispanic, and 3 Native American female faculty members in the top 50 science and engineering departments. In the top 50 computer science departments, there were no African American, Hispanic, or Native American tenured or tenure-track women faculty. With the exception of one African American full professor in astronomy, there were no female African American or Native American full professors in the physical science or engineering disciplines surveyed. Minority women are the victims of two problems: racism and sexism. The many programs that have been developed to focus on women in the past 30 years have ignored the specific concerns of women of color.

This issue has recently motivated the National Academy of Sciences (NAS) in the United States to conduct a study to determine the status of women in science in U.S. research academic institutions and to study why the representation of women in sciences and engineering is so low relative to men. The NAS Committee on Science, Engineering, and Public Policy responsible for the study appointed a group of scientists and academic leaders to the Committee on Women in Academic Science and Engineering: A Guide to Maximizing Their Potential. As a member of that committee, I have been asked to relay to the attendees of this meeting the major findings and recommendations found in that study.

The following charge was given to the committee:

1. Review and assess the research on gender issues in science and engineering, including innate differences in cognition, implicit bias, and faculty diversity.
2. Examine the institutional culture and practices in academic institutions that contribute to and discourage talented individuals from realizing their full potential as scientists and engineers.
3. Determine effective practices to ensure women doctorates have access to a wide range of career opportunities in academe and in other research settings.

4.   Determine effective practices for the recruiting and retention of women scientists and engineers in faculty positions.

5.   Develop findings and provide recommendations based on these data and other information the committee gathers to guide the faculty, deans and department chairs, academic leaders, funders, and government officials on how to maximize the potential of women science and engineering researchers.

In September 2006 the committee released its report (Committee on Maximizing the Potential of Women in Academic Science and Engineering, National Academy of Sciences, National Academy of Engineering, Institute of Medicine, 2007). Several findings in the report are described below:

•   Studies have not found any significant biological differences between men and women in performing science and mathematics that can account for the lower representation of women in academic faculty and leadership positions in science and technology fields.

•   Compared with men, women faculty members are generally paid less and promoted more slowly, receive fewer honors, and hold fewer leadership positions. These discrepancies do not appear to be based on productivity, the significance of their work, or any other performance measures.

•   Measures of success underlying performance evaluation systems are often arbitrary and frequently applied in ways that place women at a disadvantage. "Assertiveness," for example, may be viewed as a socially unacceptable trait for women but suitable for men.

•   Structural constraints and expectations built into academic institutions assume that faculty members have substantial support from their spouses. Anyone lacking the career and family support traditionally provided by a "wife" is at a serious disadvantage in academia, evidence shows. Today about 90 percent of the spouses of women science and engineering faculty are employed full-time. For the spouses of male faculty, almost half are similarly employed full-time.

The report offers a broad range of recommendations at all levels in academia. According to the report, if the committee's nearly two dozen recommendations were implemented and coordinated across public and private sectors as well as various institutions, they would improve workplace environments for all employees while strengthening the foundations of America's competitiveness. A brief overview of several recommendations from that report follows.

UNIVERSITIES

University leaders should incorporate the goal of counteracting bias against women in hiring, promotion, and treatment into campus strategic plans, the report says. Leaders, working with the monitoring body proposed by the report, should review the composition of their student enrollments and faculty ranks each year and publicize progress toward goals. According to the report, universities should also examine evaluation practices, with the goal of focusing on the quality and impact of faculty contributions. In the past decade, several universities and agencies have taken steps to increase the participation of women on faculties and their numbers in leadership positions. However, such efforts have not transformed the fields, the report says. The committee emphasized that now is the time for widespread reform.

PROFESSIONAL SOCIETIES AND
HIGHER EDUCATION ORGANIZATIONS

The American Council on Education should bring together other relevant groups, such as the Association of American Universities and the National Association of State Universities and Land-Grant Colleges, to discuss the formation of the proposed monitoring body, the report proposes. In addition, honorary societies should review their nomination and election procedures to address the underrepresentation of women in their memberships. The report also recommends that scholarly journals examine their processes for reviewing papers submitted for publication. To minimize any bias, they should consider keeping authors' identities hidden until reviews have been completed.

GOVERNMENT AGENCIES AND CONGRESS

Federal funding agencies and foundations, in collaboration with professional and scientific societies, should hold mandatory national meetings to educate university department chairs, agency program officers, and members of review panels on ways to minimize the effects of gender bias in performance evaluations, the report says. Furthermore, these agencies should come up with more ways to pay for interim technical or administrative support for researchers who are on leave because of caregiving responsibilities.

Federal enforcement agencies—including the U.S. Equal Employment Opportunity Commission (EEOC); U.S. Departments of Education, Justice, and Labor; and various federal civil rights offices—should provide technical assistance to help universities achieve diversity in their programs and employment and encourage them to meet such goals. These agencies should also regularly conduct compliance reviews at higher education institutions to make sure that federal antidiscrimination laws are being upheld, the committee said. Discrimination complaints should be promptly and thoroughly investigated. Likewise, Congress should make sure that these laws are enforced and routinely hold oversight hearings to investigate how well the Departments of Agriculture, Defense, Education, Energy, and Labor, the EEOC, and science agencies, including the National Institutes of Health, the National Science Foundation, the National Institute of

Standards and Technology, and the National Aeronautics and Space Administration are upholding relevant laws.

REFERENCE

Committee on Maximizing the Potential of Women in Academic Science and Engineering, National Academy of Sciences, National Academy of Engineering, Institute of Medicine. 2007. *Beyond Bias and Barriers: Fulfilling the Potential of Women in Academic Science and Engineering.* Washington, DC: The National Academies Press.

Trends in Basic Sciences in Contemporary Iran: Growth and Structure of Mainstream Basic Sciences

SHAPOUR ETEMAD
Iranian Institute of Philosophy

YOUSEF SOBOUTI
Institute for Advanced Studies in Basic Sciences

Iran's scientific production faced a crisis about a quarter of a century ago at the time of the revolution of 1979. The eight-year Iran-Iraq war aggravated the situation. As a result, the country experienced grave brain drain for almost a decade. However, in recent years Iran's scientific production has gained momentum. We will try to describe the dynamics and the cognitive structure of this recent growth in Iran's scientific production.

We have chosen the international Institute of Scientific Information (ISI) database for the worldwide bibliometric information of scientific papers. The journal set of this database has been expanding over the years—from 2,000 in the 1970s to 3,000 in the 1980s and to about 5,500 in the 1990s. The use of this database for studying science in developing countries is often met with reservations. For our purposes, however, it is accurate enough and serves to make comparative studies. In particular, the ISI data have provided us with sensible and sensitive information to assess the severe migration of scientists from Iran after the revolution and during the subsequent Iran-Iraq war. The scientific production of Iran, presently, is almost 50 times that in 1985, the lowest of the last 30 years. The data sample is statistically large enough to draw meaningful conclusions. The procedure was as follows:

For the period 1980–2005, the data pertaining to Iran were downloaded. Research articles, review papers, letters, and notes are, by expert opinion, the four indicators of scientific achievements. These items were retrieved from the downloaded data and analyzed. To obtain the cognitive structure, we have adopted different classification schemes for different resolution powers. For science in general, we have used Popesceau's classification. For discussing basic sciences *proper* we have used the Kirchhof–Piaget system. The latter is a comprehensive scheme, but it has minimal resolution for most of the disciplines of basic sciences.

Figure 1 is a plot of the number of ISI papers produced annually by the research community of Iran. It reflects the fact that the scientific activity of Iran has gone through substantial changes during the past four decades. If we take 1970 as our point of departure, we observe that in light of the 1973 world crisis and a noticeable increase in the country's income as a result of the increase in oil prices, Iran's performance reached a peak at the time of the revolution, 1979. Then, as a consequence of the Iran-Iraq war and the revolution, science production experienced a collapse. The performance in 1985 was reduced to almost one-fourth of the peak year, 1978–1979.

However, when one looks at the structure of the data (not presented here), one observes that the brain drain was predominantly confined to the fields of clinical medicine (the migration of specialists was mainly to North America). Therefore, when in 1988 the Graduate Study Bill (GSB) was put into action to internalize and expand the education at graduate levels, the decision was timely. The various departments of basic and engineering sciences were ready for the task. From Figure 1 we see that this decision bore fruit almost immediately. The scientific production of the country began to rise after a decade of decline and stagnation. The growth was sixfold one decade after the GSB and 40-fold by 2005.

There are, however, certain instabilities, and the dynamism is somewhat

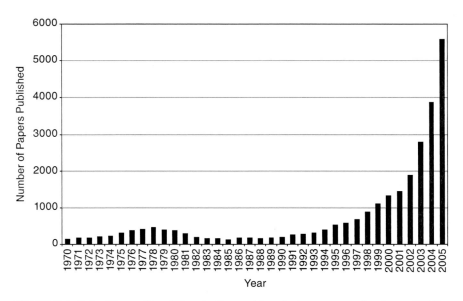

FIGURE 1 Scientific profile of Iran in the world mainstream research science (1970–2005), number of annually published ISI papers.

staggering. Before 1979, the structure of Iran's science, like that of the rest of the world, was dominated by medicine. This pattern has changed. Presently, chemistry is in the lead; notably, it is the aspect of chemistry with no significant connection to the oil industry (the major source of income of the country). Another example is the case of life sciences. In spite of considerable support by policy makers, only in 2005 does one observe a detectable growth in this field, which is, of course, welcome news. These aspects are reflected in Figure 2.

Figure 3 reflects the structure of science in 2005. For the first time in a decade, medicine takes the lead over chemistry (see also Figure 2). The world trend for publication ratios of chemistry to physics is between 2.5–2.8, and chemistry to mathematics is 3.5–4. These ratios in Figure 3 are 1.8 and 3.5, respectively. The better performance of physics should be attributed to the creation of the Institute of Theoretical Physics and Mathematics in Tehran in 1988 and the Institute for Advanced Studies in Basic Sciences (IASBS) in Zanjan in 1990. The former is a research center and the latter a graduate school dedicated to teaching and research in basic sciences.

As an indication of the research performance of the universities, Figures 4a, 4b, and 4c show the annual ISI publications of the top 10 universities per faculty member. The reason for the disproportionately better performance of IASBS resides in its dedication to graduate education. This feature at the same time demonstrates the importance of the role of graduate students, the younger generation, in the scientific development of the country.

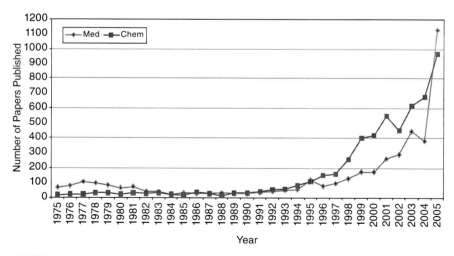

FIGURE 2 Medicine and chemistry in Iran (1975–2005), number of annually published ISI papers.

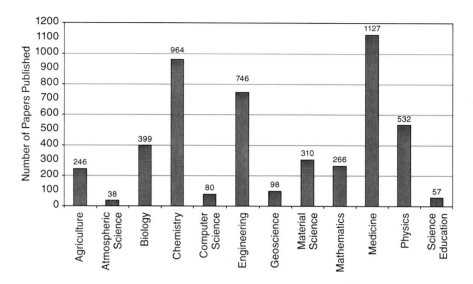

FIGURE 3 Structure of science in Iran (2005), number of annually published ISI papers.

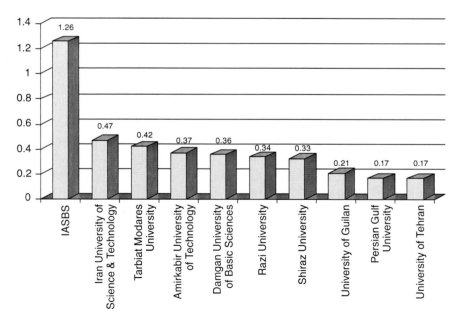

FIGURE 4a Research performance of top 10 universities of Iran, annual ISI papers per faculty member, 2002.

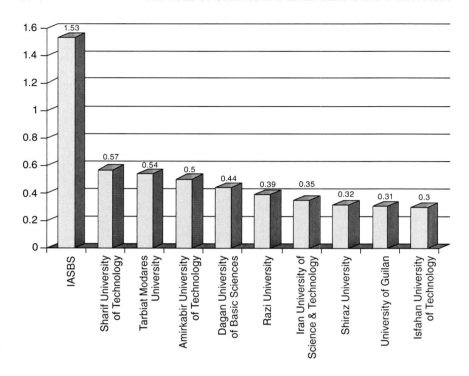

FIGURE 4b Research performance of top 10 universities of Iran, annual ISI papers per faculty member, 2003.

Figure 5 gives Iran's performance in basic sciences over the period 2001–2005. The doubling time in some cases is even shorter than five years. This raises the hope that Iran seems to have reached the threshold of self-sustainability and is capable of creating its own knowledge that leads to intensive industries and businesses.

Nevertheless, Iran's position compared with that of South Korea and Turkey shows that the country has a long road ahead. In 2005, Iran's scientific production was behind that of Turkey by a factor of 3 and behind that of South Korea by a factor of 5.5. Turkey and Iran have considerable demographic indices in common. Comparison with South Korea becomes meaningful if one considers that both are among the emerging nations on the world scene of science and technology.

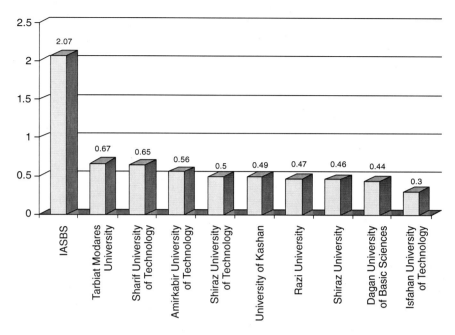

FIGURE 4c Research performance of top 10 universities of Iran, annual ISI papers per faculty member, 2004.

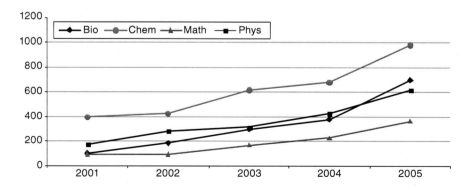

FIGURE 5 Basic science in Iran (2001–2005), number of annually published ISI papers.

ACKNOWLEDGMENTS

The data in the text and graphs are from the publications of the Ministry of Science, Research, and Technology, Office of Evaluation of Research Performances. The valuable assistance of Ms. Anahita Ashnaie and Ms. Akram Salehi in processing the data is gratefully acknowledged.

The Role of Science and Engineering in Development: Summary of Discussion

MICHAEL FISCHER
Massachusetts Institute of Technology

An overview was provided on three topics: the statistics of women in the sciences in research universities by rank and partly by field and the range of reasons behind the numbers; two organizations dealing with the continuing wide gaps between men and women in academia—the National Academies Committee on Women in Academic Science and Engineering: A Guide to Maximizing Their Potential and the Committee on the Advancement of Women Chemists (COACh); and the number of women faculty reached by COACh workshops.

The National Academies committee is chaired by Donna Shalala and composed of senior women and one man, Robert Birgeneau.[1] The committee is conducting a nine-month study to be published in fall 2007. A strategic question was raised about the wisdom of essentially only women comprising the National Academies committee, as if the issue were a women's problem rather than a societal problem. In response to a question about the number of African American women on the committee, the number indicated was three.

COACh was established in 1998 and originally targeted only chemists, but now includes other fields as well. COACh workshops have reached 350 chemistry faculty members and 600 academics from other science faculties. These workshops coach women (although men could also benefit from such coaching) on how to ask for support, how to avoid personalizing problems, and how to be

---

[1] Robert Birgeneau was MIT Dean of Science when the breakthrough report, "A Study on the Status of Women Faculty in Science at MIT" (Birgeneau, 1999) was released in 1999 and was embraced with much acclaim by MIT.

"relentlessly pleasant" in making one's needs known. COACh is planning similar workshops in Cuba, Romania, and elsewhere.

In response to a question about the essential "natural difference between men and women," one study has indicated a slight difference in three-dimensional puzzle-solving ability, a difference that disappeared if the women spent an hour ahead of the examination thinking about this type of problem. A recent study in Marseilles was also mentioned in which two groups of girls, ages 16–17, who were matched in socioeconomic and other ways, were given the same geometry problems to solve. One group was told they were mathematics problems, and the other was told that they were architecture problems. The first group scored an average of nine correct answers from 20 possible answers, whereas the second group's average score was 15. For boys the scores were the same: 13 in both groups. All of these questions were measured statistically and do not have explanatory power about particular individuals.

In addition, family issues were discussed. Assuming that tenure is granted at around age 36, there could be difficulties in waiting to achieve tenure before having a family.

Do the workshops have any demonstrable effect on promotion data? Since the program is only four years old, it is too soon to tell. At one university with a step system in rank, it was possible to develop a salary comparison of males and females in each rank, resulting in some disparities. These are public data, and so people can look up their standing and complain. Minorities were at the top of their step, but women were often underpaid.

At this university, Sunday afternoon sessions were held for women faculty on how to get information about tenure requirements, where to obtain information, and so forth. This in turn led to agitation by women for an extra year for tenure. This approach was adopted for the parent who is the main child-rearer.

A paper was presented on trends in basic science in contemporary Iran, using Institute of Scientific Information data. Among other things noted was a dramatic increase in publications in the past few years. Of interest was whether the increase in publications was a function of a new, good English language medical journal; however, there are about 10 other English language journals, including *Science from Shiraz*, the *Iranian Polymer Journal*, and so on. Ninety percent of the 5,000 papers annually are published outside Iran.

There is now a surplus of physicians being trained in Iran in response to the trend of importing physicians in the early 1980s. The ratio of university applicants to places available is about 1 to 10, which means that there are 1.4–1.6 million applicants each year.

As to support of graduate students, Ph.D. students are given stipends and sign contracts with the university to work at the discretion of the university after they finish. Fields such as chemistry, which have sufficient scientists, are no longer subsidized, and some of the contracts have even been rescinded. In Iran, higher education is essentially free. Dormitory rooms cost very little, and food is subsidized.

The distance education university Payam-e Nur was mentioned. It has 0.5 million students. Even more students, about one million, are served by the Islamic Free Universities (Azad). In both university systems, the quality is variable.

Regarding physics, and particularly astronomy, in the 1960s, Shiraz University produced a few papers. With the help of the University of Pennsylvania, it set up the Biruni Observatory. About 65 professionals, one-half faculty and one-half M.S. and Ph.D. students, produce about 15 papers a year. The second main observatory in Iran is Tusi Observatory at Tabriz University. Mashhad, Zanjan, and Babulsar have smaller observatories. The Astronomy Society of Iran holds summer camps and has huge support from high school students. It campaigned for a national observatory, which has been approved, and the site selection process is in its third year. It also produces a monthly magazine, *Nojum*, now in its fifteenth year, with a sizeable readership.

Translating the results of basic research into technology and hardware was discussed. In the West, universities and industry have grown up synergistically and have developed a culture of cooperation. The structures of Iranian universities were copied from the West by intellectuals, while the approaches of companies were copied from the West by wealthy individuals. There was little culture of cooperation. The difficulties of the past two decades have made the need for such cooperation evident, and to a limited extent the two have turned to each other. However, this collaboration is tentative and new.

As to private funding of universities, the government passed legislation about five years ago mandating that all industry must spend one percent of turnover on research and development, and a fraction of that amount must go to universities. Ten years ago, the Institute for Advanced Studies in Basic Science (IASBS) in Zanjan received all its money from the government; now 20 percent is nongovernmental funding, which is impressive, given that IASBS has no medical, agricultural, or engineering programs. Sharif University now receives 50 percent of its funds from nongovernmental sources, and Amir Kabir University receives even more nongovernmental funding.

Inquiries were made about setting up companies and relations with the oil industry. Faculty members should be full time, and they are supposed to be in their offices 40 hours a week. However, this is not always closely monitored. They are allowed to start companies with the consent of the university and on the condition that the university gets a percentage of the income in return for supplying facilities. It was suggested that the universities should exercise oversight; otherwise both faculty and the university could experience difficulties, as has been the case in the United States. One participant mentioned a policy at a U.S. university stating that no graduate student can work on anything proprietary for a dissertation thesis (although if he or she happens to discover something patentable, that is acceptable).

Inquiries were also made about petroleum engineers. There is a petroleum institute in Ahwaz, Iran. Some petroleum engineers used to train in Texas, but now they often train in Calgary because of difficulties in receiving U.S. visas. Some training in this field is offered at Amir Kabir University.

Science parks, particularly the Sheikh Bahai Science Park in Isfahan, were discussed. Sheikh Bahai Science Park was the first science park and is now one of the best developed and most successful, being more than ten years old. This has not been enough time to develop a routine and culture, however. At IASBS there is a two-year-old small information technology incubator. In 18 months it has hosted 20 companies, and the companies have transacted U.S. $1 million worth of business. There is also Pardis Science Park associated with Sharif University.

REFERENCE

Birgeneau, R. 1999. A study on the status of women faculty in science at MIT. *MIT Faculty Newsletter* XI(4). Available at: http://web.mit.edu/fnl/women/women.html. Accessed February 27, 2008.

OBSTACLES AND OPPORTUNITIES IN THE APPLICATION OF SCIENCE AND TECHNOLOGY TO DEVELOPMENT

Technology for Health: Are There Any Limits? Economic, Ethical, and Overall Societal Implications

KENNETH SHINE

The University of Texas System

In the first part of the twenty-first century, biological research will increasingly shift from reductionist approaches to the integration of knowledge. This implies the transition from genomics to proteomics to systems biology. The new "space" frontier will be neuroscience, with a focus on memory, mind, and learning. These efforts will be facilitated through the continuing application of mathematics, modeling, increasingly powerful computer capacity, and applications from physics, chemistry, materials science, and engineering to biology.

Advances in science will increasingly depend upon critical masses of scientists from multiple disciplines working together around highly focused areas. These groups will require expensive core facilities and increasingly expensive equipment and technology. Return on investment in science, including translation of basic discoveries to improve health, will become increasingly emphasized. The application of stem cell biology and genetic therapy and the explosive growth of minimally invasive procedures, such as fiber-optic surgery, have the potential to be very effective therapies. Some of the therapies will have lower unit costs but will lead to increasing societal demand. This will mean that overall health care costs will rise as the volume of procedures increases.

These developments will require increasingly difficult decisions about priorities and programs and their funding. Institutions and their scientists will increasingly seek collaborations in order to achieve the required scientific expertise and to afford core facilities and infrastructure. Increasingly, societies will be obligated to make explicit decisions about cost-effective therapies and choices about what therapies are made available and to whom.

It will be essential to focus on innovations that produce significant and substantial advances as opposed to expensive incremental approaches such as marginal modification of drugs or devices. Even in very-high-technology environments, improvements in patient outcomes often depend upon simpler interventions, such as aspirin or beta-blockers in patients with myocardial infarction.

Patients and their families will achieve increasing knowledge of science and medical options through the Internet, chat rooms, and other communication devices. This knowledge will increasingly prepare them to participate in joint patient–doctor decision making. Dr. John E. Wennberg has demonstrated that such joint decision making often leads a significant proportion of patients to choose less radical and less expensive therapies (see, e.g., O'Connor et al., 2007). Health care delivery and science in the first part of the twenty-first century is a "team sport" in which all members of the team must understand each others' goals, objectives, backgrounds, and expectations in order that science progresses and health improves.

The proliferation of scientific options will increase the pressure to make appropriate decisions for patients toward the end of life and those with poor medical prognoses. Information technology, which is essential to improved health care, also poses major problems in security and privacy. Increased incentives for the creation and application of interventions to prevent illness will be essential, not only to improve health but also to limit the rate of rise of health care costs. Preparing a health care workforce for a changing paradigm of delivering services and research will be a major challenge. Interactions with industry to achieve a translation of research to products are essential but will continue to present many moral and ethical challenges.

The range of scientific and technological opportunities and discoveries will continue to require careful ethical judgments, which should be independent of preconceived political and theological ideologies.

REFERENCE

O'Connor, A. M., J. E. Wennberg, F. Legare, H. A. Llewellyn-Thomas, B. W. Moulton, K. R. Sepucha, A. G. Sodano, and J. S. King. 2007. Toward the "tipping point": Decision aids and informed patient choice. *Health Affairs* 26(3):716–725.

Addressing Water Security:
The Role of Science and Technology

HENRY VAUX
University of California, Berkeley

Managing the world's water resources to address problems of water security will be one of the great challenges of the twenty-first century. Problems of water security typically occur on local and regional scales. Yet such problems are so pervasive that it is reasonable to think of them in global terms. Globally, the single largest problem is scarcity. In many regions, existing water supplies are inadequate to serve existing demands for municipal, industrial, agricultural, and environmental uses. This scarcity will intensify as world population increases by several billion or more over the next 30 years. Moreover, current levels of supply will almost surely be reduced because of groundwater overdraft, which is found in virtually every region of the world; declines in water quality, including soil salinization; and global climate change.

Water security can be defined at different levels. The first is the security of supplies needed for drinking, cooking, and sanitation. Although there is ample water to meet the basic personal needs of the entire population of the globe now and for the foreseeable future, a billion people lack access to good-quality drinking water and twice that number or more lack access to adequate sanitation services. This circumstance can be attributed in part to the fact that the availability of water is highly variable over the Earth's surface. Some regions do not have adequate supplies of adequate quality to provide for personal use in all seasons. The second level of security focuses on the adequacy and sustainability of water for agriculture. Agriculture is the largest consumptive user of water worldwide. Irrigated agriculture is known to be more productive than rain-fed agriculture. As the population grows, increasing numbers of countries will lack the indigenous water supplies necessary to grow the food needed to feed their own populations.

A third level of water security is water for the environment. Environmental uses of water yield environmental services such as air and water purification, production of biomass, genetic diversity, and environmental stability, as well as amenity services. Competition for water from urban and agricultural uses threatens the adequacy of supplies for environmental purposes. Developing countries with high rates of population growth are unlikely to have adequate supplies for environmental services and amenities in the future.

Globally, total water supplies are probably insufficient to provide people in every locale and region of the world with complete water security at each of these three levels all of the time. However, in many regions and locales it will be possible to do a better job of providing water security than what is being done currently. At the outset, it is important to acknowledge that there is much existing scientific information that could be helpful that is not being used. The scientific community needs to find ways of explaining, communicating, and educating that are more effective than in the past. Even significant success at this task will not obviate the need for additional science and technology. Indeed, the development and application of additional science and technology will be crucially important if food and water shortages are to be averted and the various water-based environmental services and amenities are to be protected. The remainder of this paper emphasizes the role and contributions of science in addressing the world's water problems.

SECURITY OF SUPPLIES FOR PERSONAL USE

More than one billion people currently lack access to drinking water supplies of appropriate quality. These people have no alternative but to drink contaminated water, and this results in more than 50 million deaths annually and an incidence of waterborne diseases rarely seen in the developed world. Even in developed countries, advanced water treatment technologies such as chlorination fail to provide complete protection against some pathogens, for example, cryptosporidium. More than 2 billion people lack access to adequate wastewater treatment and sanitation services. Water quality is degraded, resulting in increased incidences of waterborne diseases beyond what can be attributed to simple lack of adequate drinking water supplies. In short, many of the world's poor, particularly those in developing countries, pay horrific costs in terms of morbidity and mortality because of the absence of adequate security of water supplies for personal use (Jury and Vaux, 2005).

The problem of personal water security can be attributed to scarcity, to the lack of infrastructure to distribute water, and to a generalized failure to protect and enhance water quality. Water and wastewater treatment technologies are expensive, and the need to tailor them to site-specific circumstances makes them even more costly. Such technologies are not viable in the poorest parts of the world because of their cost. The consequence is that small-scale, decentralized technologies are and will be very important. There are a number of "soft-path" technologies that can increase the overall productivity of water at the basin level. These include treadle pumps, which are manually operated, low-cost diesel pumps, and simple technologies for treating drinking water and managing waste-water. Further efforts to refine and develop new soft-path technologies are likely to have very significant benefits (Rijsberman, 2004).

In the developed world, advances in membrane technology are about to culminate in a revolution that will allow larger volumes of water to be treated less expensively than in the past. Although seawater conversion remains too costly for most circumstances, the costs of desalinating brackish waters are now competitive in many of the arid and semiarid portions of the developed world. Bioreactors also offer the prospect of significant gains in the productivity and efficiency of wastewater treatment processes. There may be opportunities to develop applications of these new technologies that are relatively inexpensive and can be easily employed in the developing world. These applications can be realized only through additional investment in research and development, and it appears that only the developed world has the financial resources to make such investments (NRC, 2003).

SECURITY OF WATER SUPPLIES FOR AGRICULTURE

The security of water for agricultural purposes is a matter of great concern. The projected increase in population would seem to require an expansion of irri-gated acreage at a time when other factors, such as growing urban regions and urgent demands for water for the environment, indicate that agriculture may have to shrink. As previously noted, agriculture is the largest consumptive user of water worldwide and in most individual regions. The problem can be placed in stark perspective by some analyses based on the proposition that each person requires 1,500 m^3 of water annually for personal use and for growing the food needed for adequate and healthy nutrition. A common classification scheme defines countries with 1,000–1,500 m^3 per capita annually to be under water stress, those with 500–1,000 m^3 per capita annually to be experiencing water scarcity, and those with less than 500 m^3 to be experiencing extreme water scarcity (Falkenmark and Rockstrom, 2004; Zehnder, 2004).

The distribution of per capita water endowments worldwide for the year 2000 shows that there were a number of countries that had less than 1,500 m^3 per capita and were thus unable to grow all of the food needed to feed their populations. Those countries include many in the Middle East and Africa. When the effects of anticipated population growth between 2000 and 2025 are taken into account, the picture changes for the worse. Those countries with less than 1,500 m^3 by 2025 would include all of the countries on the Horn of Africa and in southern coastal Africa as well as in North Africa. Furthermore, Afghanistan, Iran, and India fall into this category, and China is likely to unless it can develop virtually all of its surface water supplies (Zehnder, 2004).

In these circumstances, typically countries that are water-short begin to import food. Since it takes some amount of water to grow food crops, importing food is equivalent to the act of importing water. This water is often called "virtual water." Statistics for Israel show how a country with inadequate water supplies can offset that inadequacy by importing virtual water. When the virtual water is counted, annual per capita supplies move toward the 1,500 m^3 figure. The evidence shows that countries with inadequate water supplies also respond by importing cereal grains. This finding is borne out for virtually every country with inadequate water (Zehnder, 2004).

This means that the intensifying world water scarcity is likely to affect the water-rich developed countries in the form of increased demands for cereal exports. The largest cereal exporters are the United States, France, Canada, Argentina, and Australia. The ability of these countries to significantly expand cereal exports is a serious issue. The western United States and Australia have their own problems of water scarcity, and it is questionable how much additional water, if any, can be made available for agricultural purposes. Although Canada is water rich, 90 percent of its waters flow north to the Arctic while 90 percent of the population lives within 100 miles of the U.S. border; and agricultural lands are, for the most part, not in the basins flowing north. The result is that the strategy of offsetting water shortages by importing virtual water may not be indefinitely expandable because water-rich countries of the Americas and Europe will ultimately be subject to their own water constraints. The ability of developing countries to generate foreign exchange with which to purchase cereal grains may be quite limited. This could also be a problem.

The need is for innovation that will make irrigated agriculture a more efficient user of water through increasing the economic productivity of water in agricultural use. The first requirement is to pair appropriate crops with appropriate growing environments (e.g., rice is not an appropriate crop for a desert environment). Second, the opportunities for improvements in on-farm water management are large and entail the use of existing information as well as the generation of new science. Advanced irrigation technology that allows growers to apply water with great precision and improvements in irrigation scheduling are among the most promising avenues. In addition, work on the moisture stressing of crops at strategic points in the life cycle or annual cycle has promise of leading to water management regimes that produce high-quality crops with little penalty in terms of quantities. The results of this work may be equally of benefit to developing and developed countries (Jury and Vaux, 2005).

ENVIRONMENTAL WATER SECURITY

The need for improving the economic productivity of water in agriculture is driven not just by the growing demand for food and fiber but also by the need to preserve some water to allocate for environmental purposes. Historically, water for the environment has been the supplier of last resort. That is, when additional water was needed for urban areas or to support irrigated agriculture, the resulting impoundments and diversions led to a decline in the quantities of water available for environmental purposes. Today, with widespread recognition of the value of environmental services such as biodiversity and water purification as well as the amenity values of aquatic and associated terrestrial ecosystems, the supplier of last resort is frequently thought of as agriculture. Nevertheless, as the global demand for food and fiber grows, it may be very difficult to find sufficient water to maintain aquatic ecosystems. Indeed, one of the dilemmas of the emerging water situation is that the need for additional water to serve agriculture may result in the loss of environmental stability and sustainability in many parts of the world. Thus, any improvement in the economic productivity of agricultural water use probably means there will be more water for the environment (Jury and Vaux, 2005).

One of the great difficulties with environmental uses of water is that so little is known about them. There is some evidence to suggest that damages to ecosystem values may not be especially large until some critical threshold is reached. More research is needed to confirm this and to identify that threshold. Interrelationships between aquatic ecosystems and terrestrial ecosystems are known to be important, but they have not been clearly identified and described. Science can also contribute to the resolution of conflicts, for example, between water development and preservation of species. It seems clear that the costs associated with underallocating water to the environment could be very high. However, much more certainty and definition will be required to confirm this supposition. Without such confirmation, learning from experience could be enormously expensive and unpleasant.

ADDITIONAL CONCERNS

Finally, there are two overarching areas that will need continuing and even increased attention from the scientific community if global water problems are to be addressed successfully. The first is global climate change. The general implications of global climate change for water supply are understood but the implications for specific regions at specific latitudes are more difficult to forecast. It does seem clear that climate change will entail increasing frequencies for extreme events such as floods and droughts. With further research the forecasting of climate change impacts should become more reliable, and more detailed information on regional and local implications will become available. Nevertheless, some uncertainty will remain and will have to be managed. As a general rule, the prescription calls for maintaining as much flexibility as possible to enhance adaptation to climate change as it occurs. This points to the importance of devising appropriate institutions and policies, and that leads to the second concern.

For the most part, current institutional arrangements for managing water were developed in different eras to achieve purposes that differ from today's primary need, which is to manage scarcity flexibly. Water management institutions around the world suffer from a tendency to focus too narrowly on the achievement of single rather than multiple water management objectives. Institutions tend to be fragmented, and there is minimal coordination among them. The consequence is a general failure to manage water holistically. Existing institutions also tend to divide concerns about water quality and water quantity artificially, contributing further to a lack of holistic management. Finally, most water institutions are not at all suited to deal with problems of scarcity. Despite the need for innovative and creative institutional arrangements, financial support of social science research related to the management of water resources has almost disappeared. Research in the social sciences is every bit as important as research in the more traditional physical and biological sciences if water problems are to be successfully addressed.

The picture that emerges from this analysis is that global water problems are now serious and are likely to become more serious as the population grows and as the economies of the world expand. Failure to use existing science and failure to invest in new science will make the problem of managing the intensifying global water scarcity difficult or impossible to address. Finally, as important as scientific research will be, improving the capacity of the scientific community to explain, communicate, and educate may be of equal importance.

REFERENCES

Falkenmark, M., and J. Rockstrom. 2004. *Balancing Water for Humans and Nature*. London: Earthscan.

Jury, W. A., and H. Vaux, Jr. 2005. The role of science in solving the world's emerging water problems. *Proceedings of the National Academy of Sciences* 102(44):15,715–15,720.

NRC (National Research Council). 2001. *Envisioning the Agenda for Water Resources Research in the Twenty-First Century*. Washington, DC: National Academy Press.

Rijsberman, F. 2004. Sanitation and access to clean water. Pp. 498–527 in *Global Crises, Global Solutions*, Bjorn Lomborg, ed. Cambridge, UK: Cambridge University Press.

Zehnder, Alexander. 2004. Feeding a more populous world. Paper presented at Sackler Colloquium on the Role of Science in Solving the Earth's Emerging Water Problems, Irvine, CA, October 8–10. Abstract available at: http://www.nasonline.org/site/PageServer?pagename=SACKLER_water_zehnder. Accessed March 30, 2008.

Obstacles and Opportunities in the Application of Science and Technology to Development: Summary of Discussion

GERALDINE RICHMOND
University of Oregon

The discussion following a presentation on technology for health care indicated that Iranians are using the Internet increasingly for researching health care topics and treatment options, but not to the level of more developed countries. In France, treatment choice typically rests with physicians; however, in the United States joint decision making between patient and physician has become the norm. In Iran, there are problems that are more immediate and basic than in the United States. For example, urban sanitary water remains an important public health concern.

As patients are given more information, they are more likely to choose a lower amount of expensive care. One great challenge in developing countries is is to focus on very simple but effective strategies, such as solar-generated energy, limited medication, and so forth, rather than focusing on spending money on large, expensive items.

In Iran, much of the nation's health care is provided by universities. The question of whether there is a strong connection between researchers and practitioners was discussed. When universities assumed responsibility for health care services, there was much uncertainty about the outcome.

In Iran, religion does not generally interfere with medical decisions. After the revolution, Eastern-style medicine was emphasized over Western-style medicine. Today this is acknowledged as a poor decision.

The discussion following a presentation on the role of science and technology in addressing problems of water security focused on water usage in Iran and the United States. Eighty percent of Iran's water is used in agriculture, primarily in rural areas with the lowest literacy rates. During the discussion a desire to change this situation was expressed. Using water more efficiently will require more education for farmers; however, very simple techniques should be employed when appropriate.

Seawater desalting is currently not economically feasible in most of the world. In Iran, there is a heavy reliance on water from mountains, where climate change is a problem. Furthermore, public awareness of climate change is not high in Iran.

One option in dealing with water scarcity is pricing the water for its scarcity value. Training for water conservation only seems to be effective during droughts, and the effectiveness is short lived. There is a lack of appreciation for this issue by many segments of the economy.

The security of water for personal use may become a problem. Distribution and simple uses are important factors to consider. There is a lack of social science research on use of water resources.

SCIENTIFIC THINKING OF
DECISION MAKERS

How to Promote Scientific Thinking Amongst Decision Makers

ALIMOHAMMAD KARDAN
Academy of Sciences of Iran

People are always faced with contradictory situations whether in their private or social life. They are forced to choose from among different alternatives and make decisions on this basis. Such choices and decision making range from trivial affairs like drinking tea or coffee to choosing a spouse, a field of study, or a job. People decide on the basis of their familiarity, interest, motive, mental and physical preparations, and general attitude. For instance, someone who sees a beverage for the first time may wish to try it; and if he already knows that beverage but does not like its taste, he will not choose it unless he does not want to disappoint someone else, such as a host.

The same applies to economic, political, and social decision making by managers within organizations and institutions. If, for instance, a manager of an economic or administrative organization is obliged to make a decision on increasing or decreasing the number of staff members, or if he decides to make changes in some of the units of his organization, the decisions he makes depend on his familiarity with the problem and its solution. Obviously, such decision making, as compared with those regarding private matters, have their own consequences, which may be of prime significance and prominence. If decision making is wrong or hasty, it may endanger the prestige of the management and the efficiency and status of the organization, or it may deprive the organization of beneficial situations. In the face of present competition, such a state is an obstacle for desirable development and expansion of the organization and may even threaten its very existence.

DEFINITION OF DECISION MAKING

As stated by management scholars, decision making can be defined as the identification and choice of a route, as taking measures toward the solution of a problem, or as benefiting from the existing opportunities.

The term *problem* refers to anything that militates against the ability of managers to achieve the stated goals of the organization. The term *opportunity* implies anything that provides an occasion for decision makers to achieve something more than those included in the stated goals.

Decision making may be preplanned. In this case, current affairs may be addressed through ordinary methods based on intuitive thinking. Yet, regarding the other kind of decision making, unplanned affairs, which are mostly related to future development, decision making is more important and more complicated. It cannot be achieved by the use of intuitive or conventional models. In fact, it necessitates the application of the scientific or rational model.

INTUITIVE AND RATIONAL MODELS

Usually, managers who are not familiar with the rational model and scientific thinking use methods and paradigms derived from a type of intuition, which is founded on personal experience and may have improper or incorrect information. In intuitive thinking, the mind relies on schemata, stereotypes, and generalizations, which for the most part do not conform to existing realities. For example, a manner of decision making that heavily relies on a particular option for all practical purposes sets aside other options. As a consequence, the use of methods based on rational models or scientific thinking is recommended rather than the use of the intuitive model. The method mentioned above uses the following steps as the best method to obtain the desired opportunity: the identification of the problem or the new opportunity, the identification of solutions to the problem or opportunity, and the choice of the best solution to the problem.

Decision making through the rational model necessitates the major steps of scientific research listed below:

- identifying or recognizing the problem (or opportunity) and specifying the factors causing the problem,
- compiling and analyzing of the information related to the problem,
- making hypotheses for the solution to the problem,
- testing hypotheses and choosing the most appropriate solution, and
- taking sensible measures for the implementation of the solution.

In addition to the identification of the problems, there may be a series of ethical issues that would make it impossible for the decision maker to put his solution into practice. These moral issues take different forms. The most important is a sense of irresponsibility and the fear of taking action.

GROUP DECISION MAKING AND ITS PROBLEM

To have a better understanding of the problem or the attainment of a desirable opportunity and to find the best solution, most managers or directors take advantage of work committees. In this way they try to clarify the problem and to find the right solution by seeking advice from experienced people. They take such measures because they generally believe that group decision making is better, more sensible, and more precise; however, research conducted by social psychologists during the 1970s shows that this is not always the case. In group discussions, a type of dominant thinking exists that is referred to as group thinking. Group thinking is more evident when there is more cohesiveness among the group members, the members have less information about external reality, and influential and dynamic personalities are present in the group.[1]

Polarization is another phenomenon in decision making that results from group thinking. That is, if the group members are in favor of or against the subject of discussion to some extent, they may go to extremes after interacting with each other and group thinking, and they may make difficult and dangerous decisions. Hence, the rational model of decision making suggests that managers should not rely too heavily on group decision making, particularly for finding a solution to a problem. If they do choose to use group decision making, they should carefully scrutinize the conclusions of the discussion or group thinking with scientific measures.

THE PROMOTION OF SCIENTIFIC
THINKING OF DECISION MAKERS

Considering the points already mentioned, one method for correct decision making when dealing with critical social, economic, and political problems is to enhance scientific thinking of mangers and decision makers. As far as the author knows, little research has been conducted in this area. Based on his knowledge and experience, the author proposes the following measures that appear to be beneficial:

1. Create the belief among decision makers that one cannot solve problems by means of conventional or intuitional methods. For appropriate decision mak-

---

[1] In this regard, Baron (1989:230) states: "When group attraction and commitment are coupled with several other factors—isolation of the group from outside information or influences, the presence of a dynamic, influential leader and high stress from external threats, an unsettling process as group-thinking may set into operation."

ing, one should identify the true nature of the problem, analyze the situation, and conduct quantitative and qualitative research before taking any further steps.

2.   In most cases, the problems are the result of a variety of factors that are beyond the specialization of management or a particular technical field. Therefore, encourage decision makers to gather general information, to have a broad view, and to avoid narrow-mindedness resulting from extreme specialization.

3.   Familiarize decision makers with the methods and skills of scientific research, especially in management and in the handling of probable crises.

4.   To avoid the drawbacks of personal experience, prejudice about that experience, and the use of only easily accessible information, accustom managers to others' criticism of their manner of thinking or actions. Here, critical thinking refers to realistic recognition of the strengths and shortcomings of any thought.

5.   Decision makers should habitually ask managers' and experts' views about the relevant problem, draw decision makers' attention to the point that they should seek the advice of others while avoiding heavy reliance on group thinking alone, and emphasize the importance of scrutinizing the views of the group and of avoiding extremes resulting from thought or decision polarization.

6.   Encourage the participation of decision makers in conferences in which the findings of scientific research about different problems and opportunities are set forth and critically discussed.

7.   Reward those who apply scientific thinking in the process of decision making to reach the right decisions and those who are endowed with a research ethic, as described in this paper.

REFERENCE

Baron, R. A. 1989. *Exploring Social Psychology*. Boston, MA: Allyn and Bacon, p. 230.

MANAGEMENT AND UTILIZATION OF SCIENTIFIC KNOWLEDGE

The Role of International Scientific and Technical Cooperation in National Economic Development

NORMAN NEUREITER
American Association for the Advancement of Science

Several assumptions underlie this presentation. First is that economic development means increasing the application of technology to raise a nation's standard of living, to free its population from a life of subsistence agriculture, to improve health and health care, and to effectively join a world commercial community that is basically driven by technology. If a nation or a people define development in some other way or with different national objectives, then the following assertions may not be valid.

If a nation is to advance by increased application of technology, it must invest in building a technological infrastructure. This has to start with capacity building, meaning the development and training of people who can adopt, adapt, and use technology obtained from abroad. This demands the construction of an educational infrastructure. At the simplest level, it can mean simply training people to use machinery and equipment that is imported from elsewhere. One example would be automobiles, which are available essentially everywhere in the world. With them has come enough technical training and experience so that they can be maintained and serviced as required.

At the most sophisticated level, a complete education infrastructure will involve primary and secondary schools that teach science and mathematics; colleges and universities where faculties in science and engineering are doing research to contribute to the world's store of basic knowledge and can train students in leading-edge science and technology; and finally a technology-based industrial community that can provide employment for graduate scientists and engineers. Partnerships and cooperative projects with faculty from other countries can accelerate the development of such infrastructure, but this requires a major, long-term policy and financial commitment by the host country.

At the university level, there are many different mechanisms for scientific and technical cooperation. One, of course, is the training of students at the undergraduate, graduate, and postdoctoral levels abroad. This training builds a cohort of scientists and engineers in both countries that know each other and find it easy to cooperate in areas of common interest after their education is finished. Another form is the undertaking of joint projects between two university laboratories, with exchanges of personnel and ideas and joint publication of results.

Cooperation in "big science" projects whereby scientists from many countries share a unique facility is important. An example is a particle accelerator to study high-energy physics or a synchrotron applied to materials and biological research; it can be either a national facility or an international facility such as the European Organization for Nuclear Research (CERN).

Still another form of cooperation is between laboratories of the ministries in two countries—such as Ministries of Health, Agriculture, or Environment.

In the corporate world, one form of cooperation is foreign direct investment (FDI) by multinational companies or companies based in other countries. Historically, such investment has often been seen by the receiving country as exploitive in that it may be focused on extraction of natural resources such as oil or minerals, with revenues benefiting only a small number of people. However, good bargaining as well as proper incentives by the receiving country can make such arrangements truly cooperative and mutually beneficial, although it is also true that for countries that have joined the World Trade Organization, the ability to require technology transfer as part of an FDI arrangement has been constrained. Another variation in FDI is the establishment of joint-venture partnerships, with a company in the host country and the investing company each contributing significantly to the joint program.

Today, all such efforts must be seen against the background of a rapidly globalizing world economy, accelerated by the instant communications provided by the Internet, but also with individual country economies vulnerable to the realities of global financial trends.

Many countries have been involved in space science projects on National Aeronautics and Space Administration (NASA) missions that have benefited their national space programs. In Europe, the high cost of space research led to the formation of a collective organization of European countries that has been a good partner of NASA on many missions. It is estimated that 15 nations have already invested $60 billion thus far in the International Space Station (ISS). Despite the cost overruns, reduced plans for science, and a serious delay in ISS construction due to the tragic loss of a space shuttle in 2003, NASA has committed to finish this project along with its international partners.

A very interesting project is the International Thermonuclear Experimental Reactor (ITER) —a project to generate energy from nuclear fusion. While work has been going on for more than 20 years on this project, the recently reconstituted international consortium of the United States, Japan, the European Union,

China, Russia, South Korea, and India is on the verge of beginning to build the reactor in France. This is a very complex program, both technically and organizationally. If it succeeds, fusion could be a long-term source of energy for an energy-hungry world, a world that is presently hostage to the caprice of global oil markets, the polluting effects of coal, and growing fears of rising carbon dioxide levels in the atmosphere.

Intellectual property is an increasingly important element in international science and technology cooperation. It is essential that the cooperating parties understand and agree how the benefits from commercialization of successful results will be divided. Particularly for a developing country entering into cooperation with a wealthier developed country or a multinational corporation, it is important to clarify the plans for ownership of intellectual property developed in the cooperative project. Intellectual property rights are also important with regard to unique native plants or other biological materials that are collected by international firms or cooperating institutions, since these natural products may contain chemical compounds with great potential value in the pharmaceutical industry.

There are a large number of problems in the world common to many countries—for instance, dealing with natural disasters such as earthquakes and floods, securing enough clean water, understanding and mitigating the effects of global warming, dealing with HIV/AIDS, treating drug abuse, achieving protection against possible pandemics such as avian flu, and so forth. It only makes sense to cooperate in these areas. To benefit from such cooperation, however, a country should have enough domestic capability to be a useful partner. These large challenges lend themselves to multilateral cooperation, and the United Nations Educational, Scientific and Cultural Organization, other UN agencies, and international scientific societies have been very useful in tackling such problems. Often a UN framework has made possible cooperation where political tensions limit bilateral cooperation.

A country with an effective national development strategy will find ways to benefit from significant developments in science and technology wherever they may occur. A nation that fails to have an appropriate international science and technology strategy will find its scientific and engineering community limited in scope and its economic development hampered. For countries of different sizes, different levels of development, and different historical traditions, international science and technology strategies will necessarily be different but nevertheless essential. It is useful to examine the role that science and technology cooperation has played in several countries.

Taiwan is an interesting example. The government focused first on agriculture, seeking to ensure that the new nation could feed itself. The government also continued the traditional Chinese emphasis on education, particularly science and engineering education. Many graduates traveled to the United States for advanced study. Many stayed to work in the United States, while others returned to Taiwan.

Next, Taiwan encouraged FDI, but rather slowly at first and with considerable effort to control what was being done by the foreign companies that invested. The government invited senior industrial managers and academicians from the United States to study the situation and make recommendations for future development. Taiwan was particularly attractive for electronics companies, which had assembly operations for semiconductors and for finished electrical and electronic equipment that required high labor input. Components and parts were shipped in from abroad, and finished products were exported largely to Western markets. The availability of large numbers of female workers from country villages and farms who were skilled at exacting assembly activities and worked for a fraction of the costs of labor in the United States and Europe made Taiwan an attractive site for Western corporate investment.

Over time, these plants became training grounds for local residents at all skill levels—in manufacturing, management, plant operation, business, finance, application of computers, international trade, and so forth. Furthermore, with government pressure, the technology level of these foreign-owned operations steadily increased in sophistication. Many young Taiwanese left the foreign factories and started their own businesses. Some began as suppliers to these plants, since a local technological infrastructure was needed in the neighborhood of these plants. Today, some 30–40 years later, Taiwan has developed its own electronics industry, fully competitive in global markets and driving U.S. and Japanese companies to higher technological levels to meet the competition. It is also very interesting that Taiwanese investment in mainland China has played an important role in China's high-tech development, despite the hostile official relationship between the two governments and the periodic threats of armed conflict.

Japan took a slightly different approach to development in its recovery from World War II. Despite the almost complete destruction of Japan's industrial base in the war, they still had a skilled workforce. Furthermore, the United States was interested in helping Japan recover from the war as a bulwark against the aggressive communist regimes in the Soviet Union and China—particularly after the Korean War in the early 1950s—and encouraged U.S. companies to enter into technology licensing arrangements with Japanese companies. Japan was reluctant to accept FDI and the establishment of foreign-owned businesses in Japan. Japan focused on rebuilding local industries by licensing foreign technology, sometimes entering into joint ventures, adopting a strict quality regimen, and stressing engineering (rather than basic science) in universities. Japan also created a special system of state-led capitalism that could finance technical and industrial development, protect domestic markets from foreign competition, and emphasize exports of manufactured products at prices often lower than those in Japan.

Cooperation of the United States in permitting trade imbalances to grow and in providing a guarantee of Japan's security under the U.S. nuclear umbrella also contributed to Japan's tremendous industrial success. However, as Japanese exports began to dominate some segments of global markets, serious trade frictions with both the United States and Europe developed in the 1980s. They lasted until increased competition from Asian countries and rising costs in Japan resulted in a recession in the mid-1990s that lasted for more than a decade. To reduce costs of manufacturing, throughout this period the Japanese continued to invest heavily abroad—in Singapore, Malaysia, Taiwan, and particularly China. Also, for graduate and postdoctoral work in the sciences and in medicine, many Japanese went to the United States. However, most returned home after completion of their studies, even though reentry into social and professional life in Japan was not always easy for the returnees.

Korea pursued a combination of the Taiwanese and Japanese models, with an emphasis on protected markets and joint ventures. Their approach was technology licensing from abroad rather than massive foreign investments alone. Success in rebuilding Korea after the devastation of the Korean War is well known. Again, an emphasis on scientific and technical education, heavy government investment in research and development facilities, concessionary funding for promising industries, and the return of many young engineers and engineering managers from the United States contributed to this success. In electronics, efforts to surpass Japan resulted in remarkable performance by Korean industry.

India's recent successes in information technology began more than 45 years ago with the establishment of the Indian Institutes of Technology through direct cooperation with leading research universities of the United States and Europe. The graduates of these excellent institutions often went to the United States or Europe to find suitable employment and today populate many U.S. companies and universities. However, a large number have returned to India and are contributing to the information technology boom. Some are working at research and software design centers that many U.S. companies have established in India in recent years.

Companies do not just blindly invest in another country. Conditions must be right to attract foreign investment. I spent nearly 25 years with Texas Instruments (TI), a semiconductor company, with operations at that time in many countries. TI had a form containing some 200 questions that had to be answered in evaluating a possible site for investment abroad. If a particular country seemed potentially attractive, the first question was always political stability. During my tenure, TI closed operations in Curaçao (labor trouble), El Salvador (after the TI plant had endured civil war for 10 years), and Argentina (uncontrolled inflation). FDI in a high-technology enterprise abroad can be a major contributor to development. However, without a responsible government ensuring political stability, reasonable economic operating conditions, and freedom from corruption, investment will seldom be made. As mentioned earlier, the host country also needs sufficient capability at the government level to understand the objectives of the investing company and to negotiate with it effectively. The host country's development interests should also be served—for example, in training people, advancing its technological capabilities, and protecting local intellectual property.

In addition to semiconductors, TI also had a national defense business. Prior to the 1979 revolution in Iran, the defense systems group was preparing for major contracts involving several different defense electronics products. However, after TI's senior vice president made a trip to Iran, he immediately ordered that all business with Iran be stopped and all TI personnel be pulled out of the country. He had found the financial situation in Iran to be so corrupt that TI simply should not do business there.

Scientists easily find a common language, even when there are serious political differences between two countries. During the darkest days of the Cold War, there were forums such as the Pugwash Conferences where U.S. and Soviet scientists, especially physicists, were able to meet. They began to talk about the massive nuclear arsenals. Tens of thousands of nuclear weapons had been built on each side and thousands were deployed and targeted at each other—in submarines, ICBM silos, and intercontinental bombers. The largest single weapon ever tested had an explosive force equivalent to 50 million tons of TNT. It was recognized by scientists from both countries that a nuclear exchange could destroy both countries and kill hundreds of millions of people. From these shared perceptions there arose a certain atmosphere of trust between the two scientific communities that was communicated to the political leaderships and eventually resulted in a series of arduous negotiations and agreements that over many years led to a remarkable degree of stability in the bipolar world of that era.

There have been other examples of how scientists have been able to find a basis for agreement, even when politicians and diplomats have found it difficult. That is why I believe that international scientific and technical cooperation can be a positive and constructive instrument of foreign policy, contributing on the one hand to national economic development when that is the agreed objective, but also serving as a mutually beneficial element of substantive engagement between two countries even in the face of political conditions that make normal diplomatic or economic intercourse impossible.

Scientists around the world do indeed speak a common language. It is often useful to let them talk.

The Role of Chemistry and Biology in the Future Development of Iran

MOJTABA SHAMSIPUR
Razi University

The publication rate of Iranian scientists in international journals has quadrupled over the past decade. More than 30 percent of publications belong to the field of chemistry alone. In fact, Iranian scientific output has skyrocketed since 1993, placing the country well ahead of most of the Islamic countries (currently second after Turkey). Meanwhile, the average impact factor of the Iranian papers has also risen considerably.

The admirable advances in the field of science and technology, brought about after the Islamic revolution of 1979 in the country, have transformed Iranian lives in a multitude of ways. This is quite evident in almost every facet of human endeavor, including the economy, health, transportation, communication, agriculture, engineering, and so forth. In the current 20-year perspective of Iran, science and technology are considered an imperative for sustainable national development, with the national goal of self-sustainment in all scientific fields.

During the past two decades, the extensive high-level research in the chemical and biological sciences together with the outstanding promotion of scientific activities by the Iranian Chemical Society and Iranian Biological Society have also strongly influenced the promotion of their interfacial sciences, including biological chemistry, biotechnology, nanoscience, and nanotechnology. During this period, the government of Iran has increased investments in support of fundamental and applied research throughout the country and especially in the fields of biotechnology and nanotechnology. According to the fourth social, economic, and cultural development plan, nanotechnology is one of Iran's priorities in technology. The government has established a special committee for nanotechnology development (National Committee for Nanotechnology), which is headed by the Deputy President and composed of representatives of six ministries and five experts and managers.

Since the mid-1980s, the country has witnessed the birth and rapid growth of several highly productive research institutes where biologists and especially chemists play key roles. They include the National Research Center for Genetics Engineering and Biotechnology (NIGEB) and the Institute of Biosciences and Biotechnology, with widespread activities in genetics and molecular biology, medical biotechnology, plant biotechnology, animal and marine biotechnology, and environmental biotechnology. The Iran Nanotechnology Laboratory Network, with 39 main member laboratories, provides researchers and active industries in nanotechnology with laboratory services and covers most of their technical needs. Most of the primary equipment relating to identification and characterization of nanostructural materials and also the equipment relating to nanoresearch in medical and biotechnology fields are available in the network. The Iran Polymer and Petrochemical Institute, the Institute of Biochemistry and Biophysics, and the Chemistry and Chemical Engineering Research Center are also among the research centers.

SCIENCE AND TECHNOLOGY STRATEGIES
FOR NATIONAL DEVELOPMENT

Technology management is a very wide-ranging topic. All countries require three groups of technologies to support national development: technologies for basic needs, technologies for the improvement of quality of life, and technologies for wealth creation. Science and technology strategy must therefore be holistic and comprehensive, addressing the role and needs of the major players in a nation, which are government, industry, the science and technology community, and society at large. The role played by each of them is briefly described:

• The government must be generally supportive, provide means for science and technology development, and use technology to improve efficiency.

• Industry must be able to use technology for creating wealth, thus enhancing economic growth. It must support science and technology development and be the major developer of technology. It must also support the generation of knowledge.

• The scientific community must make an all-out effort toward the generation and uses of scientific knowledge and innovation.

• The community at large must support a scientific infrastructure that enriches the scientific culture.

• National development perspectives must be prepared very carefully by taking into account the national vision, a long-term perspective and a five-year development plan.

• There are basically three aspects of the development of science and technology, namely policy, strategy, and action. These need to be carefully prepared and implemented.

It is well known that technology management is a key factor in modern knowledge-based productive systems. Technology management encompasses the various mechanisms, processes, and infrastructure necessary to foster, promote, and sustain the development of science and technology, the organization of science and technology activities, and putting science and technology to work for economic growth and for the attainment of overall national development objectives.

In the public sector, the various components of technology management include integration of science and technology planning into overall national development planning as well as science and technology contributions to objectives, goals, and plans. This approach needs a science and technology advisory system and a process for free flow of information between public and private sectors and between science and technology sectors and other sectors.

The various components of technology management in the private sector are

• integration of technological elements into scenario building and corporate planning;
• mechanisms for technology input and coordination, including technology surveillance and technology prospecting;
• technology acquisition and enhancement, including in-house or out-of-house sourcing at the interfirm level; and
• mechanisms for human resources development (at the firm and interfirm levels).

Figure 1 describes the flow of science and technology information in development planning, which involves the society, government, economy, and science and technology strategies and programs.

The science and technology planning process that is the key to growth is quite complicated. It involves a large number of factors, such as national vision and goals, policies, strategies, finance, evaluation, and so forth. Figure 2 describes the interrelation of various factors involved in the science and technology planning process.

For the success of the science and technology planning process, information flow between industrial development, policy planning, science and technology development, and national development policy is very important. A large number of factors and parameters need to be known and coordinated. Figure 3 gives a brief description of the interrelation between various factors and issues.

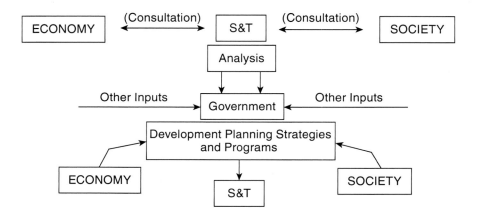

FIGURE 1 Flow of S&T information in development planning.

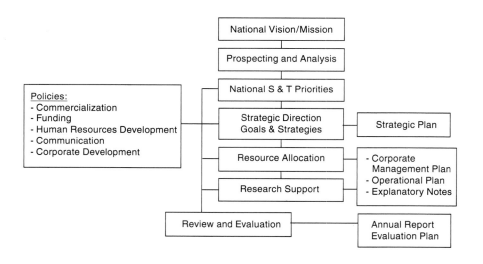

FIGURE 2 The planning process.

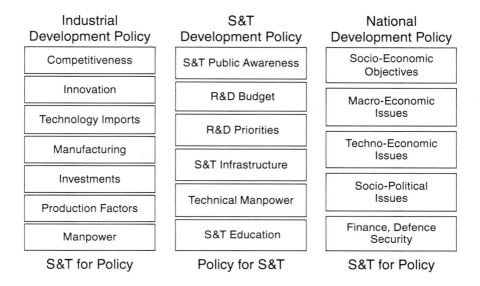

Industrial Development Policy	S&T Development Policy	National Development Policy
Competitiveness	S&T Public Awareness	Socio-Economic Objectives
Innovation	R&D Budget	Macro-Economic Issues
Technology Imports	R&D Priorities	
Manufacturing	S&T Infrastructure	Techno-Economic Issues
Investments	Technical Manpower	Socio-Political Issues
Production Factors		
Manpower	S&T Education	Finance, Defence Security
S&T for Policy	Policy for S&T	S&T for Policy

FIGURE 3 Information flow between systems: policy for S&T and S&T for policy.

PRODUCTION OF SCIENCE IN IRAN

As previously noted, since 1993, the publication rate of Iranian scientists in Institute of Scientific Information (ISI) journals has skyrocketed. Such achievements have been highlighted by well-known scientific publications. Some examples are given below.

*   Iran's long march (*Nature* 435:247–248, 2005, www.nature.com/nature

The publication rate of Iranian scientists in international journals has quadrupled over the past decade. The Iranian scientists have put together strong research groups while carefully steering clear of politics. They are determined to help build a research infrastructure that will outlast their own careers. Foreign scientists should assist their grassroots efforts.

*   An Islamic science revolution? (*Science* 309:1802–1804, 2005, www.sciencemag.org

The first fruits of Iran's biotech achievements are ripening. The Agricultural Biotechnology Research Institute has completed field trials of a genetically modified variety of local rice called *Farom molai*. Risk assessment and biosafety studies of the rice, equipped with the gene for making a *Bacillus thuringiensis* protein that is toxic to insects, are under way. At NIGEB, the plant biotechnology group is conducting field trials of various resistant sugar beets and herbicide-tolerant canola, and the industrial biotech department is scaling up in a new pilot plant for production of a recombinant human growth hormone.

•   Middle Eastern nations making their mark (*Science Watch* 14(6), 2003, www.sciencewatch.com/nov-dec2003/SIN_nov-dec_page1.htm

In recent years, Iran has substantially increased its presence in world science, according to papers indexed by Thomson ISI between 1981 and 2002. A new Science Watch survey examines the output and impact of a selected group of Middle Eastern nations over the past 20 years. As Figure 4 shows, Iran's output in science, although still comparatively small, has increased sharply in the past decade.

•   Bridges to Iran (*Lancet* 359:1960, 2002,www.thelancet.com

Iran has high educational standards. Its universities and health care system are well structured and well founded, but different. Iran prefers the medical university model to the Western model of medical faculty within a university. Iranian medical universities are responsible for health care delivery.

There is a direct correlation between the production of science and economic growth of different nations in the world. The first-ranking country in science

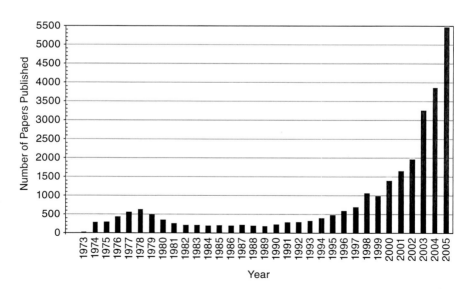

FIGURE 4 Publication rate of Iranian papers since 1974.

production is the first in economic growth. Thus, it is not surprising for Iran to have experienced rapid economic growth since 1993, which correlates well with its rapid publication rate.

During the period of 1996–2005, 15 Iranian scientists were able to publish more than 100 scientific papers each in international ISI journals. Thirteen were Iranian chemists whose names, affiliations, and number of published papers are listed in Table 1. Three are young chemists who obtained their Ph.D. degrees from Shiraz and Tehran Universities during the past decade.

Figure 5 shows those Iranian universities with more than 100 ISI papers published in 2005.

There are 18 Iranian universities whose faculty members have published between 104 and 625 papers in international journals cited by ISI in 2005.

The Iranian scientists have also had a good level of collaborative research with scientists in different countries of the world, as shown in Figure 6, including the United States, Canada, the United Kingdom, Germany, France, Japan, Australia, Italy, India, and Switzerland, with a total of more than 1,150 papers.

The field rankings for Iran in the period 1996–2005, released by the ISI, are shown in Table 2. During this time, the Iranian scientists published 19,900 papers in ISI journals with an average of 2.79 citations per paper. The field of chemistry, with a relatively high citation per paper of 3.96, is far ahead of other scientific fields such as engineering, clinical medicine, and physics.

Since the first establishment of Ph.D. programs in chemistry in 1985 at Shiraz University and later at 15 other universities and two chemistry institutes, more

TABLE 1 List of Iranian Chemists with More Than 100 Scientific Papers in ISI Journals Between 1996–2005

Name	Affiliation	No. of ISI Papers
M. Shamsipur	Razi University	337
M. M. Heravi	Azzahra University	219
I. Yavari	Tarbiat Modares University	182
M. A. Zolfigol	Bu-Ali Sina University	153
A. R. Hajipour	Isfahan University of Technology	140
N. Iranpoor	Shiraz University	136
H. Firouzabadi	Shiraz University	135
M. R. Ganjali	Tehran University	129
H. Sharghi	Shiraz University	127
S. E. Mallakpour	Isfahan University of Technology	115
I. Mohammapour-Baltork	Isfahan University	103
M. Ghasemzadeh	Chemistry and Chemical Engineering Institute of Iran	102
A. Safavi	Shiraz University	100

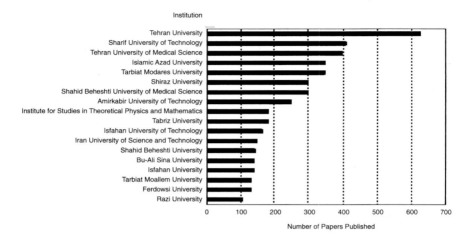

FIGURE 5 Iranian universities with more than 100 ISI papers published in 2005.

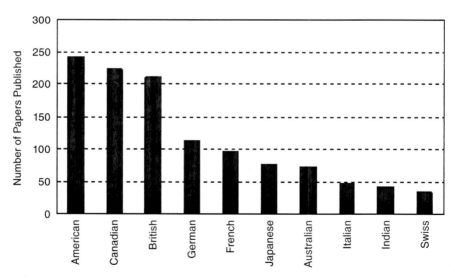

FIGURE 6 Nationalities involved in published joint research studies.

TABLE 2 Field Rankings for Iran (1996–2005)

No.	Field	Papers	Citations	Citations per Paper	% Contribution
1	Chemistry	6,100	24,176	3.96	30.6
2	Physics	1,933	7,565	3.91	9.7
3	Clinical medicine	1,986	5,480	2.76	10.0
4	Engineering	2,906	5,042	1.74	14.6
5	Pharmacology and toxicology	597	1,940	3.25	3.0
6	Plant and animal science	1,167	1,705	1.46	5.9
7	Materials science	990	1,589	1.61	5.0
8	Biology and biochemistry	565	1,345	2.38	2.8
9	Neuroscience and behavior	272	1,074	3.95	1.4
10	Agricultural science	422	801	1.90	2.1
11	Geosciences	434	744	1.71	2.2
12	Environment/ecology	291	718	2.47	1.5
13	Mathematics	783	714	0.91	3.9
14	Molecular biology and genetics	129	546	4.23	0.6
15	Immunology	143	472	3.30	0.7
16	Psychiatry/psychology	125	435	3.48	0.6
17	Computer science	463	342	0.74	2.3
18	Space science	119	253	2.13	0.6
19	Social sciences, general	148	189	1.28	0.7
20	Economics and business	34	38	1.12	0.2
	All fields	19,900	55,507	2.79	

than 450 Ph.D. students have graduated, and more than 3,500 scientific papers based on their theses have been published in international journals.

Meanwhile, Ph.D. programs in biology have also been established in the country, although at a lower rate than those of chemistry.

Examples of famous Iranian scientists are as follows:

- *Geber* (731) is well known as one of the first Iranian chemists.
- *Jahez* (776) was an Iranian biologist who published a book entitled *Animal* that described combat survival and adaptations between organisms and environments.
- *Rhazes (Razi)* (864) is known as the first chemist who discovered alcohol and sulfuric acid and their uses. He wrote *Secretes* and classified animals, plants, and soils suitable as elixirs. He also published several books on medicine, which were translated into Latin several hundred years ago.
- *Avicenna (Bu-Ali Sina)* (980) was a brilliant scientist who published many monographs on various aspects of sciences, including chemistry and biol-

ogy, and his books *Ghanon* (Law) and *Shafa* (Cure), translated into many languages, are the earliest medical texts in Europe.

•   *Alhazen (Ebne Meisam)* (1038) described for the first time the structure of the eye and compared its mechanism of working to a dark room.

•   *Heravi* (1541) and *Aghili-Khorasani* (1771) can be considered the founders of plant physiology, who presented for the first time new ideas on plant–soil relationships, the effect of light on plants, the increasing resistance of plants to the elements, mineral absorption via leaves, and the effect of organic materials on the taste of fruit.

JICS (http://www.ics-ir.org/jics/) welcomes high-quality original papers in English dealing with experimental, theoretical, and applied research related to all branches of chemistry. This scope includes the fields of analytical, inorganic, organic, and physical chemistry as well as chemical biology. Review articles discussing specific areas of chemical or biological importance are published. The journal is currently abstracted and indexed in Thomson ISI products: Science Citation Index-Expanded including the Web of Science, ISI Alerting Service, and Chemistry Citation Index; and Chemical Abstracts (Columbus, Ohio)

EXAMPLES OF IRANIAN RESEARCH AND DEVELOPMENT INSTITUTIONS

Examples of some key Iranian research and development centers that receive large contributions from Iranian scientists, especially chemists, are presented below.

National Biotechnology Committee (NBC)

To fulfill the objectives of the third national plan for social improvements, cultural developments, and economic progress, and also considering the importance of biotechnology and its increasing applications in different fields such as industry, agriculture, medicine, protection of the environment, national security, and defense, the NBC (http://ibw.nrcgeb.ac.ir/) was established by order of Iran's President on January 23, 2001. The committee works under the supervision of the Ministry of Science, Research, and Technology.

The NBC objectives are as follows:

•   enriching the scientific background of biotechnology and developing biotechnology,

•   expanding and improving the biosafety aspects of biotechnology applications with regard to Islamic and human values, and

•   increasing public and managers' knowledge of biotechnology.

Related professional commissions are Basic Science, Agriculture and Natural Resources, Medicine, Animals and Aquatic, Industry, Defense and National Security, and Biosafety and Environment.

National Research Center for Genetic Engineering and Biotechnology

NRCGEB was established in 1989 under the Ministry of Science, Research, and Technology (http://www.nrcgeb.ac.ir/). Since then, NRCGEB has been given a mandate to undertake original, state-of-the-art research activities. It has been established with the dual purposes of promoting research in avant-garde areas of biological sciences and biotechnology and providing advanced training and education programs for scientists and students from other universities and academic institutions.

The center activities have been focused on five major areas, including medical biotechnology, plant biotechnology, animal and marine biotechnology, industrial and environmental biotechnology, and basic sciences. In each division, a strong emphasis is placed on a highly innovative and pioneering investigation in the fields of both basic and applied biology. About 170 employees including 93 researchers (41 Ph.D.s) are currently working at the center. In addition, there are also researchers from universities and research centers from inside the country and around the globe who are cooperating with the center through shared research projects and applied educational courses.

The *Iranian Journal of Biotechnology* publishes original scientific research papers in broad areas of biotechnology. The main areas include agriculture, animal and marine sciences, basic sciences, bioinformatics, biosafety and bioethics, environment, industry and mining, and medical science. The language of this journal is English, and it is indexed by ISI.

Agricultural Biotechnology Research Institute of Iran (ABRII)

ABRII (http://www.abrii.ac.ir) is one of the national agricultural research institutes under the supervision of Agricultural Research and Education Organization (AREO). AREO is an organization of the Ministry Jihad-e Agriculture. ABRII was established in 1983 as the Plant Biotechnology Department of the Seed and Plant Improvement Institute. It was upgraded to the level of an independent institute in 1999.

ABRII's research activities are mainly focused on the field of advanced plant molecular biology and biotechnology. As a national research institute, ABRII strives to promote applied and basic research in plant science and technology and organizes workshops for advanced training of experts who are involved in research and development.

Other related biotechnology institutes include

- Pasteur Institute of Iran (http://www.pasteur.ac.ir/),
- Razi Vaccine and Serum Institute (http://www.rvsi.ac.ir/), and
- Agricultural Biotechnology Research Institute (http://www.abrii.ac.ir/).

Polymer and Petrochemical Institute

The Iran Polymer and Petrochemical Institute (IPPI: http://www. iranpolymerinstitute.org) was established in 1986 on 22 acres of land to carry out applied research in polymer science and technology and to help local industry in training engineering students at the postgraduate level, as well as taking on direct challenges to meet their needs.

IPPI, with more than 110 highly trained staff at the Ph.D. and master's levels, has eight research divisions, including rubber, plastic, composite and paint, adhesive and resin, novel drug delivery systems, polymer science, special polymers, and biomaterials. These divisions are supported by the technical and engineering department of the institute for their development projects.

IPPI departments include

- Department of Polymer Science and Catalysts,
- Polymerization Engineering Department,
- Color, Resin, and Surface Coatings Department,
- Composites and Adhesives Research Department,
- Plastic Research Department,
- Rubber Research Department,
- Novel Drug Delivery Systems Research Department,
- Department of Polymer Science and Catalysts,
- Biomaterials Research Department,
- Polyurethane Department, and
- Gas Conversion Group.

The *Iranian Polymer Journal* (IPJ, http://journal.ippi.ac.ir) includes IPPI scientific achievements. The journal is well received in international academic circles. The submission of articles from worldwide sources is a promising sign that the IPJ, within its short life, has gained international recognition.

Institute of Biochemistry and Biophysics (IBB)

IBB (www.ibb.ut.ac.ir) was founded in 1976 with the triple purposes of fulfilling research in the areas of biochemistry, biophysics, and related areas; performing collaborative research activities with domestic as well as reputable research centers around the world; and training students at the master's and Ph.D.

levels. In that year, the institute admitted 19 M.Sc. and four Ph.D. students in molecular cell biology. Ever since, and especially from 1988 onward, IBB has been actively engaged in training graduate and doctoral students in biochemistry and biophysics. The other main activity of IBB is basic and applied research in biochemistry and biophysics, including exchanging scientific information with other research centers and institutions both inside and outside the country. IBB is regarded as one of the prestigious research institutes affiliated with the University of Tehran, and as such it enjoys a high position among the research centers in the country.

Two important research centers of IBB are the Iran Bioinformatics Center and the Biomaterial Research Center.

Chemistry and Chemical Engineering Institute of Iran (CCEII)

CCEII (http://www.cceii.org/en/index.php) is housed in a four-story research building in the northwest of Tehran. This research center is affiliated with the Ministry of Science, Research, and Technology. The joint efforts of chemists and chemical engineers are to fulfill the following goals: to provide fundamental and educational research facilities related to all chemical and chemical engineering areas and to diversify the talent of chemists and engineers. To make modern facilities available at pilot and semi-industrial levels, the center has been taking an active part in recognizing, elaborating, and conducting some of the national research projects since 1989. More than 50 full-time staff members work in the research, administration, and service departments. In addition, a number of scientists and researchers from other universities and educational centers in Iran and abroad cooperate with the center in research projects. CCEII is comprised of five research laboratories in the areas of analytical chemistry, inorganic chemistry, organic chemistry, organosilicon chemistry, and chemical engineering.

Iranian Nanotechnology Initiative (http://www.irannano.org/)

According to the fourth social, economic, and cultural development plan, nanotechnology is one of Iran's priorities in technology, and there is a special committee for nanotechnology development (National Committee of Nanotechnology).

The missions of the committee include

- ratification of the goals, strategies, major policies, and plans to develop nanotechnology in country,
- classification of the responsibilities of the governmental entities, determination of their missions, and coordination among them, and
- monitoring of the implementation of the plans.

Nanotechnology Policy Studies Committee, Technology Cooperation Office

The Nanotechnology Policy Studies Committee (http://www.tco.gov.ir/nano/) has been established to study the different aspects of this technology, its applications, and its impacts on other technologies.

To formulate the country's macro infrastructure, it is necessary to have a survey of the following:

- applications and impacts of nanotechnology,
- potential of different countries in nanotechnology,
- necessary infrastructures such as national laboratories, educational programs, and scientific networks, and
- the potential in the universities and research centers as a basis for allocating research activities to them.

Atomic Energy Organization of Iran (AEOI)

The research division of AEOI (http://www.aeoi.org.ir/) consists of several research centers as follows:

- Nuclear Fusion Research Center,
- Gamma Irradiation Center,
- Center for Renewable Energy Development,
- Nuclear Research Center for Agriculture and Medicine,
- Information and Data Processing Center,
- Reactor Research and Operation Department,
- Research Center for Lasers and Their Applications,
- Yazd Radiation Processing Center,
- Bonab Research Center, and
- Nuclear Fuel Production Division.

Management and Utilization of Scientific Knowledge: Summary of Discussion

HENRY VAUX
University of California, Berkeley

A presentation on the role of international scientific cooperation in national economic development suggested that economic development means increasing the application of technology and having a process for doing so. The purposes of economic development are improving health and health care, raising the standard of living, and joining a world commercial community that is driven by technology. If science is not contributing to technology, at least in the long term, it is not significantly helping economic development. Capacity building is critical, and it can occur at different levels. International cooperation can take the form of technical assistance programs or specialists from different countries collaborating together.

EXAMPLES FROM INDIVIDUAL COUNTRIES

Taiwan started its science and technology effort by focusing on agriculture and becoming self-sufficient in food production. Taiwan also focused on education and used an advisory committee from the United States for advice in the field of education. The committee's recommendations were particularly important in the development of electronics.

In Singapore, foreign direct investment has been very important, although in many countries the history of foreign direct investment is dismal. To attract foreign investment, appropriate policies and people who can manage the investment process are needed. Managers in Singapore made it clear that they did not want only physical facilities. They wanted relationships involving schools and universities and training for middle managers and others as well. As Singapore built its value chain, education improved.

The value chain and education must advance in parallel, as is also happening in India. The Indian Institutes of Technology trained students who came to the United States and other countries and were very successful. Many have now returned to India and are working in major company research and development efforts in software and other aspects of electronics.

In Japan, after World War II, universities focused on engineering. Domestic markets were kept closed, and products were imported until they could be manufactured in Japan. Until the United States worked out a way of cooperating with Japanese companies, progress was slow. Committees were created to determine how products could be manufactured for export to the United States. Trade complications remain, however, because Japan is still interested in protecting its market.

Korea forged relationships involving companies, brought engineers back to Korea, and put the engineers on production lines to develop Korean capabilities. The engineers on processing lines soon understood how to tweak the process.

In China, science and technology policy played a significant part in opening relations with the West. Henry Kissinger hoped to give China the prospect of something more than just a political relationship. He wanted to embrace scientific cooperation. A number of initiatives were presented to China, and several of them came to pass. There was an early visit to the United States by a Chinese scientific delegation. During the Carter administration, Frank Press (Science Advisor to the President of the United States from 1977 to 1980) developed a science agreement with China. During the next 25 years, more than 50,000 students from China were in the United States almost every year. Most were enrolled in science and technology fields. China is now on an explosive path of scientific and technological development, much like India.

International organizations (such as the World Health Organization, the World Meteorological Organization, and the United Nations Education, Science, and Cultural Organization) are also an important element for economic development. The International Institute for Applied Systems Analysis in Vienna was created in 1972 to enhance peaceful relations with Russia. Now its focus is shifting toward working with developing countries on global problems.

There is a great deal of exchange between Americans and Iranians, including scientific cooperation.

One participant asked for two or three reasons why development based on science and technology has not occurred in the Middle East outside Israel—for example, in Egypt. Reasons include lack of political stability, weak infrastructure, and inadequate education, all of which hamper the development of the labor force. Iran and Turkey are advancing reasonably well. Of course, Iran has had episodes of instability, such as the revolution in the 1970s and the coup in the 1950s.

The Republic of Ireland had a problem with mass out-migration, and the agriculture system was primitive, particularly in the west of Ireland due to inheritance laws. With the advent of the European Union (EU), the Irish government embraced EU concepts, obtained substantial help, and used it intelligently. Ireland strengthened education, adopted an open-door investment policy, and for a time was the world's largest exporter of software. Political will is crucial for this type of development. In Ireland both political parties embraced the need for modernization. Still, policies have not changed the infrastructure, which may not be necessary in the end.[1]

Successful international collaboration will often require that intellectual property be shared.

The presentations on the role of chemistry and biology in the future development of Iran underscored the fact that a comprehensive science and technology strategy should address the roles and needs of the major organizations that are involved. The government should be supportive of scientific development. Industry should be able to share technology. The science and technology community should make an all-out effort to build innovation capacity. In brief, the priorities of technology should meet basic needs, raise the quality of life, and create wealth. An industrial development policy, a science and technology development policy, and an economic development policy should be integrated to advance technological development. At the same time, it is important to recognize that global and regional trade agreements can conflict with national industrial development policies.

The current circumstances in Iran show an increase in scientific activity. The number of published papers increased by more than an order of magnitude between 1986 and 2004.

Overall, the journals that publish them have a greater-than-average impact factor. According to the Institute of Scientific Information (ISI), the number of publications correlates well with economic growth in Iran. Many of Iran's Ph.D. programs can compete with Ph.D. programs abroad for developing faculty members. There are 13 professors of chemistry who each published more than 100 papers in ISI journals between 1996 and 2005. There are 17 Iranian universities that each had more than 100 papers published in ISI journals in 2005.

Additional evidence of the high level of activity and quality of Iranian scientists can be seen through the activities of the following organizations and activities:

- the Iranian Chemical Society, which has a good journal;
- the Iranian Biotechnology Committee, which is active in advancing the interests of biotechnology, with a number of ministers included as members by law;
- the Iranian Institute of Biochemistry and Biophysics;
- the Iranian Institute of Polymers and Petrochemicals;

---

[1]See also Ó Riain (2004).

- the Iranian Nanotechnology Initiative; and
- the Atomic Energy Organization of Iran.

REFERENCE

Ó Riain, S. 2004. *The Politics of High Tech Growth: Developmental Network States in the Global Economy*. Cambridge, UK: Cambridge University Press.

SCIENCE, SOCIETY, AND EDUCATION

About the Relation of
School Teachers with Science

YVES QUÉRÉ

Interacademy Panel on International Issues

The organization of school teaching in France is set in such a way that the elementary administrative cell, called a *Circonscription*, includes approximately 350–400 teachers in the same geographical area. The head of the Circonscription is an *Inspecteur de l'Education Nationale* (IEN) who, in particular, organizes one-day meetings of a pedagogical nature for the teachers. I have been invited more than 60 times in the past few years to participate in these meetings with essentially the following structure: one or two lectures about science in the morning, each 45 minutes, followed by a 1- to 1.5-hour period of open discussion. They take place in large cities as well as in the suburbs and rural areas. Therefore, I have "met" more than 20,000 school teachers, plus a number of IENs; I have heard a great deal on how they feel about science and about science teaching.

Although there would be some danger to generalize these feelings to the whole population of teachers (approximately 340,000 in all), a first impression may be drawn from those reactions with which most agree. First, school teachers in France have an initial training of three years in university, ending with a *Licence* (in any subject of their choice, such as French, foreign language, history, science, arts, and so forth, although only 15 percent have chosen science or mathematics), plus two years in a specialized institute devoted mostly to pedagogy. This means that a majority of the reactions quoted herein are from teachers with minimal training in science.

The comments heard most often are in response to the following questions: Is science easy or difficult? Open or closed? Good or bad? Necessary for development, or useless? The first question is the most frequent. The answer is almost always the same. Science is *difficult* and, in fact, *too difficult* to be taught:

"Science is definitely too difficult for me," is a sentence I have heard hundreds of times.[1]

TOO DIFFICULT?

The specific difficulty of science as seen by the teachers—not only in France, but in most countries where I have visited schools—seems to come from what they hear about science from TV and newspapers in dealing with big events such as the inauguration of a large accelerator in Geneva or a huge telescope in Chile, the launching of a man into space, or the determination of an immense genome of some plant. All these famous technical achievements give the public the idea that science is, from now on, more or less out of the normal world. If we tell them about growing a bean and measuring its growth in millimeters over time as a function of illumination or temperature, they frequently answer, "But this is not science, this is gardening!" They have the impression that the word *science* cannot cover something as commonplace and ordinary as the growth of a bean or the melting of an ice cube.

We have to explain to teachers that science is indeed like a high mountain (Everest) that only a few very specialized people (perhaps Nobel Prize or Fields Medal winners) are able to climb. Nevertheless, everybody can walk on those nice hills around the Alps, the Rocky Mountains, or the Elbourz. Everybody can find a suitable altitude.

Science provides the same possibility to reach a given level, in complete continuity and according to one's talents and training. Contrary to the fear of many school teachers, there is no initial gap that must be overcome before entering into science. One must just want to take a walk with the students and enjoy it. This is why they should not be frightened—for the erroneous reason that "it is too difficult for me"—of teaching science. This is also why scientists should leave their laboratories—their ivory towers—from time to time to visit teachers, give them examples of simple experiments to be performed and to be understood by children, and explain that these are real aspects of science.

OPEN OR CLOSED?

There was indeed a time, especially at the end of the nineteenth century, when scientific inquiry was considered to be practically finished. For physicists such as Biot or Kelvin, physics was on the verge of being completed; this naïve belief still exists in some places. It is clear now to any scientist that, on the con-

---

[1] The same teacher who finds science "too difficult" usually considers that history is comparatively quite easy. If the remark is made to a teacher that historians still discuss the immensely difficult question of the causes of the first World War, he/she will answer immediately, "We do not have to go into details with children," without noticing that the same could be said for science.

trary, each discovery opens unforeseen new fields of research. Teachers should know that knowledge is like a precious nectar poured into an amphora, each new drop increasing the contents, but at the same time enlarging the container, in a kind of endless adventure. Science is an open field of knowledge that will probably never be contained in a closed horizon. Children should be taught early that they have in their hands a huge book full of white pages waiting to be filled in by young generations.

GOOD OR BAD?

Does science spread the evil (weapons, pollution, arrogance) in societies and individuals more than the good (health care, communications)? The question—unthinkable a few decades ago when science was believed to be the source of indefinite happiness—is nowadays a matter of discussion, especially in industrialized countries. Many teachers are sensitive to these societal questions, and some may discourage children from entering the field of science when they become older on the basis of ethical concerns. Children should be told at early ages that science, and, more generally, knowledge, is immensely beneficial for humankind, but also that citizens should be aware of potential and sometimes real dangers that it may create. They should be ready to discuss these matters.

NECESSARY FOR DEVELOPMENT?

Is science useful for development? Any child should know that it is not only *useful*, it is *necessary*. There could not be an applied science without a vivid and imaginative basic science. Development is not possible (e.g., in health, engineering, agriculture) without an applied science-based technology upstream. A complete continuity exists between fundamental research and the most practical applications: no mobile phones without the *useful* band theory of solids and no treatment of infectious diseases without *useful* research on the complex geometry of proteins.

Individual and societal capacity cannot be built without a minimum of education in science and technology. Education will be proper and profitable only if it is provided by teachers who not only convey some knowledge to their students but who are themselves comfortable with science, aware of its ethical issues, and convinced of its importance, its openness, its universality, and its intrinsic beauty.

Promotion of Health Education
in Primary Schools

BÉATRICE DESCAMPS-LATSCHA
La main à la pâte

Evidence has accumulated that health education is not the province of only physicians, nurses, and other health professionals; rather, it has to be integrated with science education. Indeed, both health teaching content (how our body functions, how diseases appear and propagate, and how we can prevent them) and methodology (to observe, to propose hypotheses, to verify them, to deduce a behavior, and to evaluate its results) are scientific. Also, health problems are strikingly different between northern and southern parts of the world and strongly depend on environmental and socioeconomic conditions.

In industrialized countries, health education of children has become an important task for the Ministries of Health and the Ministries of Education. This theme has recently been introduced in school programs. Indeed, it is conceivable that health education of children may improve public health, notably in the prevention of risks and the protection against aggressions and addictions. Likewise, the sensitization of school-age children toward diseases and disabilities—avoiding stigmatization—could be of great importance for solidarity, respect, and tolerance toward "others," not only within but also beyond the school.

In developing countries, a large part of disease prevention already relies on health education of children through nongovernmental organization initiatives such as health-promoting schools and the Child-to-Child Program initiated by the World Health Organization (WHO) and the United Nations Children's Fund (UNICEF), respectively. Interestingly, these programs are based on the principle of making children both actors and messengers of health for themselves, other children, their families, and the community.

MUCH TO LEARN FROM DEVELOPING COUNTRIES

During the past 15 years, organizations such as WHO, the United Nations Educational, Scientific and Cultural Organization (UNESCO), UNICEF, and the World Bank have joined their efforts for developing health promotion strategies through health-promoting schools. Several world summit forums have been held, leading to declarations for improving the health, education, and development of children, and, through them, families and community members. The discussions have been mainly devoted to developing countries and aimed at preventing diseases related to poor environmental conditions (lack of drinkable water, malnutrition, defective personal and community hygiene) and to those pathogens responsible for pandemic infections (malaria, schistosomiasis, and HIV). Although most of these health problems are not encountered with a high prevalence and acuity in industrialized countries, most declarations were cast on a worldwide basis as indicated by the three following extracts:

• The World Declaration on the Survival, Protection and Development of Children[1] concluded that "Together, our nations have the means and the knowledge to protect the lives and to diminish enormously the suffering of children, to promote the full development of their human potential and to make them aware of their needs, rights and opportunities." Of course, this is dependent on a new opportunity to make respect for children's rights and welfare truly universal.

• The declaration at the World Education Forum advocated, "All children must be given the chance to find their identity and realize their worth in a safe and supportive environment, through families and other care-givers. They must be prepared for responsible life in a free society. They should, from their early years, be encouraged to participate in the cultural life of their societies."[2]

• This same forum stated that girls must be given equal treatment and opportunities from the very beginning. This is noticeable not only for avoiding gender discrimination but also in its support of public health. Indeed, as recently revisited by the Women Health Education Programme,[3] women's illiteracy is still the major factor involved in the abnormally high perinatal maternal and child mortality in the developing world.

---

[1]Agreed to at the World Summit for Children, UNICEF 1990. See http://www.unicef.org/wsc/declare.htm. Accessed September 13, 2007.

[2]See http://www.unesco.org/education/efa/ed_for_all/dakfram_eng.shtml. Accessed September 13, 2007.

[3]Developed by André Capron at the French Academy of Sciences through the Interacademy Panel (IAP). See http://www.whep.info/. Accessed September 13, 2007.

WHY FOCUS ON SCHOOLS?

Ensuring that children are healthy and able to learn is, without a doubt, an essential component of an effective education. Likewise, school is a key setting. There, it is possible to improve and sustain the health, nutrition, and education of children previously beyond reach (e.g., girls).

A large amount of research has shown that health and education are inseparable. Studies have indicated, for example, that cognitive performance is affected by a child's nutritional status and that illness from parasite infection results in absenteeism, school failure, and dropouts. Likewise, for education systems to be effective, efforts to increase enrollment, build more schools, or train more teachers are not enough and come to naught if physical or mental health problems prevent children from attending school regularly and remaining in school for a sufficient number of years. In its final report, the World Education Forum also stressed, "Effective health education responds to a new need, increases the efficacy of other investments in child development, ensures better educational outcome, achieves greater social equity and is a highly cost effective strategy." The report proposed the FRESH (Focusing Resources on Effective School Health) framework as a start for enhancing the quality and equity of education.[4]

THE CHILD-TO-CHILD EDUCATIONAL APPROACH

Introduced in the early 1970s, the Child-to-Child educational approach has now become a central core[5] of a worldwide movement of health and education workers to protect and preserve the health of communities through the education of children. It mostly relies on the principle of encouraging and enabling school-age children to play an active and responsible role in the health and development of themselves, other children, and their families. A great number of activity sheets and reading resources have been produced that deal with the major health topics encountered in these countries, such as breastfeeding, children's growth and development, nutrition, personal and community hygiene, diarrhea, vaccination, maternity without risks, and the prevention and cure of diseases such as malaria and AIDS.

Boxes 1 and 2 retrace, as an example, the Child-to-Child framework set up for preventing HIV/AIDS pandemic and supporting children facing the impact of this new plague for our humanity.

---

[4]See http://www.unesco.org/education/wef/en-docs/findings/rapport%20final%20e.pdf. Accessed February 13, 2008.

[5]The Child to Child Trust is currently chaired by Dr. Patricia Pridmore at the Institutes of Education and Child Health, London University.

BOX 1
Steps in the Child-to-Child Approach for Preventing HIV/AIDS

1. **Understanding the issue:** What happens to children when parents are sick with AIDS or have died?
2. **Finding out more:** How does HIV/AIDS affect their community?
3. **Discussing findings and planning action:** Children prepare posters and compose poems, songs, and dramas.
4. **Taking action:** Children organize a special event for other children and adults aimed at tackling discrimination, ensuring that all children are included, and motivating families.
5. **Doing better:** Children use all opportunities individually and as a group to help each other and younger children to cope with the impact of HIV/AIDS.

BOX 2
Support of Children Facing the Impact of
HIV/AIDS in Their Families

1. **Training** for older children on issues of growth and development of themselves and of younger children.
2. **Fostering relationships** between older children and younger orphans and other vulnerable children living with HIV/AIDS and/or having been abandoned.
3. **Counseling for children** heading households on their psychological and emotional needs.
4. **Working with parents and children** to develop memory books to help them in the future to get a sense of their identity and know they have been loved.

LA MAIN À LA PÂTE:
A NOVEL APPROACH TO HEALTH EDUCATION

The recent development of inquiry-based science teaching owes much to the efforts of the scientific community. In France, three physicists—Georges Charpak, Pierre Léna, and Yves Quéré—launched their own version of inquiry-based science teaching under the label La main à la pâte[6] (translated as "Hands on" in English). It has now spread to more than 40 percent of French schools and to many schools in numerous other countries, including Afghanistan, Argentina,

---

[6]Recently revisited in Charpak et al. (2005) and Quéré and Jasmin (2005).

BOX 3
The Ten Principles of La main à la pâte

The teaching approach:
1. Children observe an object or a phenomenon in the real, perceptible world around them and experiment with it.
2. During their investigations, pupils argue and reason, pooling and discussing their ideas and results and building on their knowledge, since manual activity alone is insufficient.
3. The activities suggested by the teacher are organized in sequence for learning in stages. The activities are covered by the program and leave much to pupil self-reliance.
4. A minimum schedule of 2 hours per week is devoted to the same theme for several weeks. Continuity of activities and teaching methods is ensured throughout the entire period of schooling.
5. Each child keeps an experiment logbook, in which the children make notes in their own words.
6. The prime objective is the gradual acquisition by pupils of scientific concepts and operating techniques, with consolidation through written and oral expression.

Partnership:
7. The family and community are solicited for work done in class.
8. At the local level, scientific partners (universities, etc.) support classwork by making their skills and knowledge available.
9. Teaching colleges in the vicinity give teachers the benefit of their experience.
10. Teachers are able to obtain teaching modules, ideas for activities, and replies to queries via the Internet. They can also take part in a dialogue with colleagues, training officers, and scientists.

Brazil, Cambodia, Chile, China, Columbia, Egypt, Iran, Malaysia, Mexico, Senegal, and Togo.

The general idea of La main à la pâte is to enable children to participate in the discovery of natural objects and phenomena, to bring them into contact with the phenomena in their reality (outside virtual reconstructions) directly through observation and experimentation, to stimulate their imagination, to broaden their minds and to improve their command of the language (see Box 3).

A number of distinct resources, including an Internet Web site,[7] are now available. The golden rule is that teachers should be trained as children will be taught.

---

[7]www.inrp.fr/lamap.

The widespread development of La main à la pâte throughout the world, under distinct terms such as Hands-on in the United States, Learning by Doing in China, or Penser avec les mains in Switzerland, illustrates well how science education at an early age could serve as a bridge across cultures. Interestingly, an international portal has recently been added to the French Académie des Sciences site by the International Council for Science and the Interacademy Panel in direct link with La main à la pâte.[8]

The La main à la pâte team has recently set up a working group to develop teaching modules of health education on major themes of public health.

Among these, the prevention of solar exposure was chosen first, because it is responsible for the high prevalence of cutaneous cancers and the increased frequency of cataracts. This theme has led the group[9] to build a teaching module, Vivre avec le Soleil ("To Live with the Sun"; see Wilgenbus et al., 2005), aimed at the sensitization of children to the characteristics of sun exposure and of its potential damaging effects as well as to the importance of simple protective measures against ultraviolet exposure. It is also hoped that, as is already happening with the Eratosthenes[10] model, by joining children from several parts of the world to measure the Earth's radius, To Live with the Sun might serve as an accurate challenge for the idea of education as a bridge across cultures all over the world.

In the past two years our group has focused on poor nutrition, a worldwide health problem that, depending on the part of the world and even on the region in some countries (e.g., Argentina), results in malnutrition or obesity. As in many other industrialized countries, obesity appears to follow an epidemic progression from very early in childhood. Recent statistics on overweight children in the United States indicating that it affects nearly 25 percent of children under the age of ten justifies the sensitization of children on the important roles of balanced diets and physical activity. In France, a threefold increase in overweight children—from five percent in 1980 to 16 percent in 2000, among which four percent are obese, has also been observed and has justified, in 2002, the initiation of a national program (Programme National Nutrition Santé) aimed at disseminating overweight prevention messages to the general French population, based on eating well and regular physical exercise.

In most industrialized countries, obesity has thus become a major public health problem, mainly through its associated complications, such as hypertension, cerebrovascular and cardiac incidents, and diabetes, which reduces the overall life expectancy of the obese by 10 years. Being overweight is a condi-

---

[8]Contact: marc.jamous@inr.fr.

[9]With the contribution of the Association Sécurité Solaire.

[10]In this project entitled "Following in the Footsteps of Eratosthenes," more than 300 English-speaking pupils have measured the Earth's circumference from their classroom, simply by observing the shadow of a vertical stick at noon local solar time. Again this time, schools of many countries will join together to reproduce the observations of the Greek scientist who, more than 2,000 years ago, was the first to propose a simple method to measure our planet's size.

tion involving many factors, including not only nutritional and physical factors but also psychological, socioeconomical, cultural, and genetic factors. Most of these are outside the scope of the health education of children. However, many countries have started to develop nutritional education programs directly within schools.

The nutrition education module To Eat and to Move for My Health recently proposed by La main à la pâte (Bense et al., 2008) is devoted to 5- to 7-year-old primary school children. It stresses the importance for the child to have both a diversified and well-balanced diet and a regular physical activity program. Following closely the La main à la pâte principles (see Box 3), it proposes a series of pedagogical sequences[11] in which each child has the opportunity to propose a hypothesis, to experiment (alone or in small groups), and to write observations and conclusions in an experiment notebook.

However, special attention has to be drawn to two major aspects of nutrition that have been at the center of our reflections when preparing that module. The first is to avoid the stigmatization of the obese child, who is often present in the classroom and has led us to propose a sequence named: All Similar or All Different? Indeed, it has to be kept in mind—and teachers themselves do not always know—that the mental suffering of the obese child is intense and often hidden behind behavioral troubles. Moreover, the obese child is an easy target of mockery and injurious jokes from other children. Together, these lead to feelings of disgust toward their own bodies and to depression.

The second deals with the absolute necessity of informing parents of the teaching procedure, obtaining their consent and, whenever possible, having them participate. Even if eating is a biological function, to eat and to give to eat are also vehicles that touch the intimate, self-representation, and family life. Nutrition education allows the development of the capacities to judge and to make social and individual changes without putting into question important habits and familial traditions.

Finally, the statement of Yves Quéré (2008) is important: "When I was a child, health education was already taught but often within The moral lesson and on a distinct day from The science lesson [...]. Science and health were two worlds which at school ignored each other." Indeed, this La main à la pâte pedagogical approach—that is, an investigation developed by the children under the direction of their teacher, initiated by their questioning, and relying on their own experimentation—herein proposed for health education in primary school contributes to demonstrating that this has to be considered as a part of science teaching.

More globally, attempts made to promote health education at primary school

---

[11]Four sequences (comprising three 1.5-hour pedagogical sessions) are entitled as follows: Sequence 1, To Move, but for What?; Sequence 2, Hygiene, Is It Important?; Sequence 3, Eating Well, but How?; Sequence 4, Drinking, But not No Matter What?

and aimed at enabling children to face present public health dangers should always incorporate the first ethical principle of Kant, "First do no harm," and its corollary, "So act as to treat humanity, whether in thine own person or in that of any other, in every case as an end withal, never as a means only."

REFERENCES

Bense, D., B. Descamps-Latscha, and D. Pol. 2008. *Manger, bouger, pour ma santé.* Paris: Hatier.

Charpak, G., P. Léna, and Y. Quéré. 2005. *L'enfant et la science. L'aventure de la main à la pâte.* Paris: Odile Jacob.

Quéré, Y. 2008. Foreword. In *Manger, bouger, pour ma santé,* D. Bense, B. Descamps-Latscha, and D. Pol. Paris: Hatier.

Quéré, Y., and D. Jasmin. 2005. When learning science becomes child play. *A World of Science* 3(3).

Wilgenbus, D., P. Cesarini, and D. Bense. 2005. *Vivre avec le soleil: activités Cycle 3: guide de l'enseignant.* Paris: Hatier.

Science, Society, and Education: Summary of Discussion

BARBARA SCHAAL
Washington University

HENRY VAUX
University of California, Berkeley

" Science is feared because it is dangerous." Science is highly admired, but an overwhelming number of school children think that science may be dangerous for humankind. There is some evidence, however, that science is not feared when it is taught decently.

"Science is feared because it is too difficult." Many teachers believe that science begins somewhere above them. They do not perceive that it is a continuum, with simple science at the bottom. Ten years ago it was discovered that only three percent of children had science in elementary school.

The mode adopted is usually vertical teaching, but now horizontal teaching is being promoted. Horizontal teaching tends to be inquiry driven as follows:

- The child asks a question.
- The teacher does not answer but sends back another question and encourages the child to pose a hypothesis.
- The children then work on a small-scale experiment.
- Children are consistently required to express themselves orally to some degree and more so in writing. An important aspect is to teach the children to write clearly.

This type of horizontal teaching has started to spread throughout the world. The horizontal teaching method teaches science at a very simple level, but it teaches the scientific method in a completely valid way. It moves away from the problem created by perceiving science as only "big science," which is incomprehensible to most people.

Conditions for the horizontal teaching method are to provide teachers with resources, to create links between teachers so that they consult with each other, and to create links between teachers and the world of science. The most important element of this method is to train the teachers as the children will be taught. This approach minimizes or eliminates the fear of difficulty. Goals for the future of this program are to have a worldwide citizenry capable of recognizing the fear of science as dangerous and to have a more focused method for addressing teachers' fears that science is difficult.

The discussion following the presentations illuminated the fact that science teaching at the secondary level is inadequate. One participant described a number of the structural barriers that need to be overcome to achieve reform in teaching science at the secondary level.

During the discussion on teaching science, society, and ethics, one participant was intrigued by the idea that scientists can be outside an experiment. Of course, many scientists may be involved in an experiment. They then interact directly with the system and influence it.

Xenotransplantation also came up. With xenotransplantation there is the risk of a pandemic, and there are immunological barriers. For example, the first heart transplants were unsuccessful because of the patients' autoimmune response and rejection. Xenotransplantation was stopped when immunological difficulties were encountered. When these were resolved, it began again. At a conference of the Institute of Medicine on xenotransplantation, the conclusion was that the field can police itself. This raises question of who polices science. Stating that self-regulation is better than having politicians regulate science brings up the same point that has been discussed before institutional review boards; namely, that the institutions should self-regulate.

For studies of populations (as opposed to individuals), ethical questions are often raised. In Iceland, genetic data were requested for the whole population. Who gives consent for an entire population? How does one receive informed consent?

Another participant commented that someone needs to be a pioneer regarding transplants. In terms of the ethics of organ donors, regulating money provided to donors will not necessarily solve the ethics problem. In addition, there is an animal rights issue associated with xenotransplantation.

Ethics is a vast field. Self-regulation does not work for stem cell research. Permission to take personal data in Iceland was rammed through Parliament. Ethicists were hired by companies in order to push their agendas. In the 1960s, James Neel made decisions about blood samples and vaccinations in the Yanomamo community. In the 1990s, ethics issues were no longer left to the scientists.

A discussion followed the presentation on the promotion of health education in primary schools. Without information for parents, the La main à la pâte health education program would not work. Parents give their consent, and children pass on messages to parents. Evaluation of the program involves an assessment of the consequences of the program.

One participant underscored that the evaluation of the program is extremely important. An example was given of an instance in which young schoolchildren were given instant food to bring home for lunch. The girls' food was taken away and given to the boys and their fathers. The solution was to serve lunch at school so that each child received an equal portion of lunch.

In the United States, obesity is becoming more prevalent. In Brownsville, Texas, 30 percent of the population is diabetic. Also, there are vending machines in many U.S. schools serving regular (as opposed to diet) soft drinks. One participant emphasized the importance of putting only low-calorie sodas in all U.S. schools. In terms of obesity prevention in the United States, a lesson can be learned from programs in place for asthma that have resulted in greater school attendance. With better funding, better programs will result.

One participant brought up the correlation between overweight parents and overweight children. He suggested the 4-H principle in which children bring home relevant information to their parents. The situation is somewhat different in Iran, where many rural areas still need education opportunities.

The La main à la pâte program may be exported to other countries, even though it is not yet implemented throughout France.

APPENDIXES

A

Workshop Agenda

International Workshop on Science and Technology and the Future Development of Societies

JUNE 27–30, 2006
LES TREILLES, FRANCE

Welcoming Remarks
Yves Quéré, Académie des Sciences, France, and Interacademy Panel on International Issues

Session 1: Science and Society Issues
Chair: Kenneth Shine

The Role of Communications and Scientific Thinking
Barbara Schaal, Washington University

Knowledge, Validation, and Transfer: Science, Communication, and Economic Development
John Enderby, Institute of Physics

The Morality of Exact Sciences
Yousef Sobouti, Institute for Advanced Studies in Basic Sciences, Zanjan

Session 2: The Role of Science and Engineering in Development
  Chair: Henry Vaux

Women in Academic Science and Engineering in the United States: Challenges and Opportunities
  Geraldine Richmond, University of Oregon

Trends in Basic Sciences in Contemporary Iran: Growth and Structure of Mainstream Basic Sciences
  Yousef Sobouti, Institute for Advanced Studies in Basic Sciences, Zanjan

Session 3: Obstacles and Opportunities in the Application of Science and Technology to Development
  Chair: John Enderby

Technology for Health: Are There Any Limits? Economic, Ethical, and Overall Societal Implications
  Kenneth Shine, The University of Texas System

Addressing Water Security: The Role of Science and Technology
  Henry Vaux, University of California, Berkeley

Session 4: Scientific Thinking of Decision Makers
  Chair: Norman Neureiter

How to Promote Scientific Thinking Amongst Decision Makers
  Alimohammad Kardan, Academy of Sciences of Iran

Session 5: Management and Utilization of Scientific Knowledge
  Chair: Barbara Schaal

The Role of International Scientific and Technical Cooperation in National Economic Development
  Norman Neureiter, American Association for the Advancement of Science

The Role of Chemistry and Biology in the Future Development of Iran
  Mojtaba Shamsipur, Razi University

Session 6: Science, Society, and Education
Chair: Yousef Sobouti

About the Relation of School Teachers with Science
Yves Quéré, Interacademy Panel on International Issues

Teaching Courses in Science, Technology, and Society and the Importance
of Ethics
Michael Fischer, Massachusetts Institute of Technology

Promotion of Health Education in Primary Schools
Béatrice Descamps-Latscha, La main à la pâte

Session 7: Summary of Workshop Sessions
Chair: Yves Quéré

Summary of Session 1, Science and Society Issues
Norman Neureiter, American Association for the Advancement of Science

Summary of Session 2, The Role of Science and Engineering in Development
Michael Fischer, Massachusetts Institute of Technology

Summary of Session 3, Obstacles and Opportunities in the Application of
Science and Technology to Development
Geraldine Richmond, University of Oregon

Summary of Session 4, Scientific Thinking and Decision Makers
Geraldine Richmond, University of Oregon

Summary of Session 5, Management and Utilization of Scientific Knowledge
Henry Vaux, University of California, Berkeley

Summary of Session 6, Science, Society, and Education
Barbara Schaal, Washington University
Henry Vaux, University of California, Berkeley

Session 8: Interacademy Programs

Reports on Past Programs
Alimohammad Kardan, Academy of Sciences of Iran
Henry Vaux, University of California, Berkeley
Michael Fischer, Massachusetts Institute of Technology
Norman Neureiter, American Association for the Advancement of Science

Multilateral Approaches
Yves Quéré, Académie des Sciences, France, and Interacademy Panel on
International Issues

Developments in Washington that Affect Programs
Glenn Schweitzer, The National Academies

Developments in Tehran that Affect Programs
Alimohammad Kardan, Academy of Sciences of Iran

Future Opportunities for Cooperation
Discussion led by Kenneth Shine, The University of Texas System

B

Workshop Participants

Béatrice Descamps-Latscha
La main à la pâte, Paris

John Enderby
Former Vice President, Royal Society
President of the Institute of Physics
Chairman, Melys Diagnostics Ltd.

Michael Fischer
Professor of Anthropology and
    Science and Technology Studies
Program in Science, Technology, and
    Society
Massachusetts Institute of Technology

Alimohammad Kardan (deceased)
Member
Academy of Sciences of Iran
Professor of Psychology and
    Educational Sciences
University of Tehran

Norman Neureiter
Director
Center for Science, Technology, and
    Security Policy
American Association for the
    Advancement of Science

Yves Quéré
Cochair
Interacademy Panel on International
    Issues
Académie des Sciences, France

Geraldine Richmond
Richard M. and Patricia H. Noyes
    Professor of Chemistry
University of Oregon

Barbara Schaal
Vice President
National Academy of Sciences
Spencer T. Olin Professor of Biology
Washington University

Mojtaba Shamsipur
Associate Member
Academy of Sciences of Iran
Professor of Chemistry
University of Razi, Kermanshah

Kenneth I. Shine
Executive Vice Chancellor for Health
    Affairs
The University of Texas System

Yousef Sobouti
Member
Academy of Sciences of Iran
Professor of Physics and Director
Institute for Advanced Studies in
    Basic Sciences, Zanjan

Henry J. Vaux, Jr.
Professor Emeritus
Department of Agriculture and
    Resource Economics
University of California, Berkeley

National Academies Staff

Glenn E. Schweitzer
Director
Office for Central Europe and Eurasia

A. Chelsea Sharber
Senior Program Associate
Office for Central Europe and Eurasia

C

Science, Technology, and Society— The Tightening Circle

GEORGE BUGLIARELLO
Polytechnic University

Since the beginning of civilization, the twin innate human quests of understanding nature in its physical, biological, and social aspect (what has come to be called science) and of modifying nature and building artifacts (the vast activity encompassing endeavors such as engineering, medicine, and agriculture, which we call technology) have had a fundamental impact on the evolution of human societies. They have also been indissolubly interconnected because to modify nature, we must understand it and to understand it, often we must manipulate it and build artifacts.[1] It would be impossible within the confines of this paper to encompass the vastness, complexity, and, as Thomas Hughes (2004) put it, the messiness of the interactions among science, technology, and society; but we can attempt to adopt a systematic framework for addressing them and to exemplify some salient points.[2]

Repeatedly, the two unstoppable quests to understand and modify nature have changed societal views, have expanded the human reach, and have fundamentally transformed society, from the discovery of agriculture and metals to the industrial and information revolutions, to today's biotechnological revolution.

---

[1] At times science has preceded technology and vice versa, often with large time lags. William Thomson, first Baron Kelvin, who established the understanding of the existence of absolute zero temperature, insisted there were no conflicts between science and technology, practicing both as an academic and as a technological entrepreneur, helping get the first trans-Atlantic cable in place (Lindley, 2004).

[2] The impacts assume at times transcendent qualities, as in the case of the Chinese invention of paper being highly regarded by early Muslims because it was being used for writing the words of God.

THE DOUBLE CIRCLE OF INTERACTIONS

The two quests have affected the very core of societal beliefs, from cosmogony to the origins of life, have determined the fates of societies and nations, and have been impacted, in turn, by society, in a double circle of interactions (Figure C-1). The possibilities offered by engineering modifications of life are changing our very understanding of life and, for the first time, are providing us with the ability to modify life. The hard-fought acceptance of the concept of verifiable truth—the bedrock of science and technology—has pervaded many aspects of societies, from law to medicine to economics to education, offering a compass to guide an ever more complex world (making the issue of truth, in essence, also one of utility).

Science and technology have transformed society's views about the future, from the need to preserve our finite resources to the attempt to avert or mitigate the consequences of cataclysms previously believed to be acts of God and from pandemics to the glimmer of hope that one day we might be able to deflect some catastrophic asteroid hits.

Equally immense but not always fully recognized is the influence of society on science and technology. The birth of astronomy responded to earlier religious needs for precise information about the movement of celestial bodies. Discoveries spurred by the school of Henry of Lancaster (Henry the Navigator) required the most extreme science and technology of the time, from mapping to nautical instruments to the caravels.

Religious dogmas and political and social ideologies as well as different philosophies have in various periods exerted a determining influence on the course of science and technology (Pool, 1997), all the way to today's stem cell

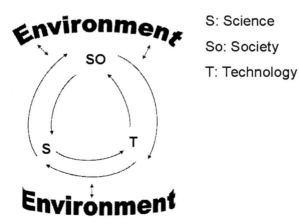

S: Science

So: Society

T: Technology

FIGURE C-1 The double circle.

controversy. Clearly, the polarization between science and religion weakens societies and continues to be unresolved (Silver, 2006).

The Greek philosophers believed in the fundamental unity and order of the world, leading them to scientific conclusions more from reasoning than from detailed observation. Greco-Roman law inspired a large segment of Islamic law. The five-element Yin and Yang theories of the Chinese helped the development of their scientific thinking, but, as Needham (2000) pointed out, their dominance simply went on too long so that China had no renaissance or reformation. Also, different philosophies affected the relation between time and science, with, for example, time for the ancient Chinese being one-directional and for the Hindus cyclical. The Chinese conception of harmony and the Hindus conception of the interpenetration of observation and observer (in a sense a forerunner of Heisenberg's uncertainty principle) have had profound impact on their science, as had, for the ancient Greeks, the conception of symmetry and balance. In the Koran, the concept of knowledge (*ilm*), the second-most recurring word, inspired inquiries in the areas of science, philosophy, and technology and led to respecting and preserving the texts of the ancient Greeks. As Needham observes, there is little to choose from between European and ancient Chinese philosophy with regard to the foundation of scientific thought, with Western scientific minds concerned with "what essentially is it?" and Chinese ones with balance; two equally important conceptions. In Europe in the seventeenth century, formal logic retarded independent thinking and became clearly inadequate to handle change. Thus, philosophical views set the stage for scientific and technological developments but can also channel and confine them. As Cyril Smith (1979) insightfully stated, if somewhat categorically, "discovery, from its very nature must at first be illogical, unforeseen and outside the framework that previously exists. . . . In moving beyond what is already known and well understood, logical thought is of less value than the complex reaction of the entire human being."

The Renaissance, with its new spirit of inquiry and discovery, opened the gates to new scientific and technological developments, and later, the French Revolution affirmed the separation between science and religion. Thanks to this new spirit, transportation networks that were the sinews of the Roman Empire, but had disappeared in the medieval fracturing, could be succeeded more than a millennium later in different form by the transportation networks of ships, roads, and railroads that supported the expansion of the British Empire and today's global networks of telecommunications and aviation. The late-nineteenth-century creation of the research university became a mechanism for systematizing and accelerating the process of discovery, as first triggered by von Liebig's laboratory in Germany. The industrialized battlefield in World War I and its exponential technological growth in World War II led to an unprecedented mobilization of science and technology and to the birth of formalized science and technology policies. And one cannot overlook the seminal influence of art over the millennia on the development of science and technology, from the use of metals to Leonardo, and from the 1909 futurist manifesto to today's new art media.

Thus, society's political systems, culture, and organizations are the essential ingredients for the creation and enhancement of science and technology. Only well-organized societies are able to build large public works and logistic networks. The culture of Japan made possible its rapid modernization after 1854; today global corporations, financial institutions, and venture capital have become key enablers of discoveries and technological development.

In general, the culture of the nineteenth century encouraged a great flourishing of science and technology, which in turn led to the modernistic culture of the twentieth century. However, society is not a monolith. Scientific and technological developments may impact certain aspects or parts of society faster or differently from others, whether one considers laws, the attitudes of leaders, military prowess, commerce, health, or education. At times the developments are handicapped by the short vision of society's leadership, as when Napoleon missed the importance of the steamboat or the British Admiralty before World War I that of the submarine. Furthermore, there can be vast societal and scientific underestimates of the time to future developments, as when Aldous Huxley in 1931 predicted that human spaceflight would not occur before 2970 (Barrett, 1990) or when Wilbur Wright reportedly said to Orville in 1901, two years before their first successful flight in 1903: "Man will not fly for fifty years."

THE TIGHTENING OF THE CIRCLE

The circle of reciprocal impacts of science, technology, and society and their propagation around the world have greatly accelerated in the past 100 years and even more in the past few decades, as exemplified by the spread of automobiles and cell phones, by the diffusion of new forms of entertainment, by rapid changes in the global economy, or by the proliferation of nuclear weapons since 1945. The interaction of the science–technology–society circle with the environment is also tightening, as is evident from global warming or the deadly failures of societal responses to the 2004 Indian Ocean tsunami and to Katrina. The circle is tightening not only in the interaction of science, technology, and society but also within each of its components—within science, for example, in the biological field with its impact on other areas of science; within technology, where innovations build on innovations; and within society, with accelerated changes in art, in our views of history, and archeology (Gere, 2005) and in concepts of human needs and rights.

A significant factor in the tightening of the circle is technology transfer, a sociotechnological process that began when we emerged as a distinct artifact-creating species and the knowledge of our inventions spread from neighbor to neighbor. Today, more than ever before, the future of nations, economies, industry, health care, education, and other key human activities is profoundly affected by the effectiveness of the complex sociotechnological aspects of that process and of national systems of innovation.

Although accelerating, the circles of science, technology, and society interactions often have played out over very long periods of time and may continue to do so. Some 2,500 years elapsed from the Phoenician invention of glass to that of optics and the telescope; another 500 years elapsed before the invention, in rapid succession, of lasers and fiber optics, with their expanding societal impacts from communications to medicine. Similarly, 2,500 years separate the philosophical atomistic conception of Demokritus from Bohr, Seaborg, Fermi, and the subsequent nuclear military and civilian applications. Rocketry, probably an early Chinese invention and used extensively in India in the late 1700s by Tippu Sahib, eventually made possible intercontinental missiles, Sputnik in 1957, and the 1969 Moon landing, with geopolitical impacts from militarization of space to the concept of "Spaceship Earth," to new international laws defining the limits of national sovereignty in space.

Information is yet another example of recent acceleration of a long cycle, with a 1,600-year gap separating the Library of Alexandria from Gutenberg and the Reformation, 500 years separating Gutenberg from the first Apple computer, 22 years from the Apple to the Internet, and only four years from the Internet to the World Wide Web in 1993.

CONSEQUENCES

The speed of scientific and technological developments is straining the ability of society to adjust to them, as in the case of globalization; to control them, as in the case of nuclear weapons proliferation; and to take full advantage of the opportunities they offer, such as those of genomics and genetic engineering.

The greater the speed, the greater is the penalty for being left behind and being unable to innovate, as happened to the Ottoman Empire after the seventeenth century, to China between 1500 and 1950, and more recently to the former Soviet Union.

The greater the speed, the greater are the possible unintended consequences. The diffusion of the automobile, particularly in the United States, led to extended suburbs and has created the enormous geopolitical pressures we are experiencing today because of the need for oil. Industrialization is leading to global warming; information technology led, through the radio, to women's active political participation in the United States and today is giving us unfathomable cyber vulnerabilities, the weakening of national sovereignty, and the centralization of diplomacy and the conduct of war (Bugliarello, 1996). Worrisome job dislocations result from globalization, and genetic engineering is provoking religious backlash and raising concerns about discrimination in insurance and in the workplace. We are still trying to determine the possible unintended consequences of nanotechnology and genetically modified foods (Ross, 2006). The ever-tightening of the circle exacerbates the problem in that new technological inventions outpace by necessity the crucible of evolution that operates in the biological world, even if the extinction patterns of biological species and business firms might be guided by the same laws of failure (Kean, 2006; Ormerod, 2006).

FUNDAMENTAL QUESTIONS

Three fundamental questions arise from the circle of science–technology–society interactions: *questions of science*, that is, how can we go about understanding nature, and what are the limits to our understanding (for instance, is the origin of the universe knowable?); *questions of engineering and technology*, that is, how can we go about modifying nature; and *questions of societal ethics*, that is, how far should we go in doing so (Figure C-2)?

These three fundamental questions lead to more specific derivative ones, such as: Science for whom? If something can be built, should it be built (the question of technological determinism, as in the case of cloning or of weapons of mass destruction)? What is the nature of the contract between science and society? Should there be one, explicitly or implicitly? What policies should guide it? (The control of science and technology by society is easy enough. The trick is how to leave room for new ideas, discoveries, and innovations.) What are the societal settings necessary for changes of scientific paradigms and technological innovations? Why do some societies develop through sophisticated science and technology while others stagnate?

The questions of expectations and of progress loom large in the public understanding of science and technology, often leading to exaggerated expectation of further developments (Sigma Xi, 1993). A facet of the problem is an increasing blurring today of the boundary between superstition and science, which has sci-

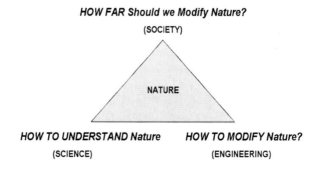

FIGURE C-2 Fundamental questions in the interaction of science, engineering, and society.

entists spending much time in attempting to promote scientific and technological literacy and in re-explaining established scientific principles to a general public that is not rationally informed. Unfortunately, the task is made more difficult because many branches of science and technology have become hermetic to the outside, and education has, by and large, failed to convey to the general public the criticality of the efforts of fact-finding and the anguish of decision-making in creating new technologies and in modifying nature. The question whether science and technology represent real progress inevitably arose after the hecatombs of World Wars I and II and also persists because of the enduring poverty of billions. Closely connected to these questions are those of the relation of science and technology to religion, ethics, freedom, and government, crystallized by the debates about stem cells or creationism and about the relation of scientific truth to other truths. The relation of government to science and technology is complex (Morgan and Peha, 2003): Democracy encouraged the development of Western science, but nondemocratic societies such as the Chinese, the Persians, the Hindus, and the Muslims also developed impressive science. In certain respects, this occurred even with some absolute forms of government, but the intrinsic lack of systems of checks and balances inevitably led to wrong turns, as in the direction of biology in the Soviet Union under Lysenko or the case of Galileo under papal absolutism.

These questions and issues need to be revisited every generation and with every major new scientific, technological, and social development. The ability to do so is crucial to our future, a future that the ever-tightening double circle of interactions of society, science, and technology thrusts upon us with ever greater speed and consequences.

URGENT ISSUES

At the beginning of this century, a host of very urgent issues confronts us, as discussed in a recent publication (Bugliarello and Schillinger, 2004). Even a partial list underscores their criticality: (a) global sustainability in the light of the eco-environmental impact of the unprecedented growth of human populations and their exploding concentration in cities, which now encompass about one-half of the world's population; (b) the need for a new paradigm for the role of human work in society as the dream of delegating work to machines begins to be realized, with humans being increasingly displaced by machines and automation and not any longer automatically definable through their jobs; (c) the seemingly irreducible problem of poverty, with still more than one-fifth of the human race below the poverty level; (d) our reliance on some technological systems so complex as to make it impossible, with potential catastrophic consequences, to completely predict their performance; the mitigation of disasters, made ever more dangerous by the concentration of populations in areas at risk; (f) the changing face of war, now more precise in its ability to reach physical targets, but also potentially more lethal; (g) the needed revolution in health care because of genetic engineering and other technologies and also because of the enormously pressing questions of affordability and of the protection against pandemics, now enabled to migrate at unprecedented speed; (h) the sustainability of food and water resources; (i) the emergence of biomachines, combining attributes of both biological organisms and machines; and (j) the prospect of creating artificial life.

The complexity of these issues demands ever more urgently a societal ethic to define what we should or should not do. Approaches and answers are different across cultures, but they need to be harmonized if a world endangered by the actions of humans and by natural cataclysms is to have a human future. Our species must endeavor to agree as to the fundamental issues it faces, forming, as it were, a superhighway of fundamental principles receiving inputs from various local ethics. That is, can our field of view be expanded to possibly identify ethical and behavioral aspects to which, consciously or not, all cultures subscribe?

The ultimate responsibility of science and technology to society—and vice versa—is to give wings to moral force and to do the utmost for the sustainability of our world and the survival of our species.

We are now in what could be the midpoint in the existence of the Earth, from its genesis five billion years ago to its possible end five billion years from now (Bugliarello, 2000). If up to this point natural selection and speciation have dominated and humans have adapted to the environment, the future of our species will be determined by the concurrent evolution in the social, biological, and machine domains. This "biosoma" evolution is the result of the ever-tightening circle of interactions of science, technology, and society, in all their complexity, their dangers, and their promise. It demands, if it is to have a future, the fostering of a new way of thinking.

REFERENCES

Barrett, D. B. 1990. Chronology of futurism and the future. In *Encyclopedia of the Future*, G. T. Kurlan and G. T. T. Molitor, eds. New York: Simon & Schuster, Macmillan.

Bugliarello, G. 1996. Telecommunications, politics, economics, and national sovereignty: A new game. *Technology in Society–An International Journal* 18(4):403–418.

Bugliarello, G. 2000. The Biosoma: The synthesis of biology, machines, and society. *Bulletin of Science, Technology & Society* 20:452–464.

Bugliarello, G., and G. Schillinger, eds. 2004. Technology and science entering the 21st century. Special Issue, *Technology in Society—An International Journal*: 26:99–536.

Gere, K. 2005. *The Tomb of Agamemnon.* Cambridge, MA: Harvard University Press.

Hughes, T. P. 2004. *Human-Built World. How to Think About Technology and Culture.* Chicago, IL: University of Chicago Press.

Kean, S. 2006. Nothing succeeds like failure. *Science* 312:531.

Lindley, D. 2004. *Degrees Kelvin: A Tale of Genius, Invention & Tragedy.* Washington, DC: Joseph Henry Press.

Morgan, M. G., and J. M. Peha, eds. 2003. *Science and Technological Advice for Congress.* Washington, DC: Resources for the Future.

Needham, J. 2000. *Science and Civilization in China.* Cambridge, UK: Cambridge University Press.

Ormerod, P. 2006. *Why Most Things Fail—Evolution, Extinction and Economics*. New York: Pantheon.

Pool, R. 1997. *Beyond Engineering—How Society Shapes Technology*. Oxford, UK and New York: Oxford University Press.

Ross, P. E. 2006. Tiny toxins? *Technology Review* 108(2). Available at: http://www.technologyreview.com/read_article.aspx?id=16814&ch=nanotech. Accessed March 31, 2008.

Sigma Xi, The Scientific Research Society, 1993. *Ethics, Values and the Promise of Science, Forum Proceedings*. Research Triangle Park, NC: Sigma Xi.

Silver, L. M. 2006. *Challenging Nature: The Clash of Science and Spirituality at the New Frontiers of Life*. New York: Ecco, HarperCollins.

Smith, C. S. 1979. Remarks on the discovery of technique and on sources for the study of their history. In *The History and Philosophy of Technology*, G. Bugliarello and D. B. Doner, eds. Chicago: University of Illinois Press.

D

Current Issues on the Utilization
of Scientific Findings

HASSAN ZOHOOR
Academy of Sciences of Iran

The utilization of information, knowledge, and science in the processes of decision-making, policy-making, and planning is an issue that every country in the world faces. Research, as the most essential process for the production of new knowledge, has been for a long time of prime significance both in developing and developed nations. Each society makes every effort to advance science as the most developed stage of knowledge and to exchange and utilize science for increasing knowledge. Each country strives to identify methods for the improvement of cooperation between researchers and decision makers and consequently for increasing the effectiveness of decisions made for comprehensive development. This is the traditional concept of the role of research and development.

The aim of the author is not to provide a list of issues and obstacles but to analyze the types of relationships between scientific findings, decisions, and development planning as well as to provide strategies for the improvement of such relationships.

To define the role that science can play in the process of development and to introduce methods for the improvement of such a role and for strengthening of the links between scientific findings, decisions, and policies, we should review some of the issues and limitations. It is important to study the utilization of scientific findings in decision-making and strategies for its improvement as well as to trace the role of science and scientific findings in plans and programs. In this way, we can discuss the utilization of scientific findings without any nonscientific bias and arrive at a reasonable approach for laying the groundwork for utilization of scientific findings.

One of the best approaches is to define benchmarks for the level of science utilization. Then the attainment level could be measured according to those benchmarks in each society and even in each economic sector of a society.

UTILIZATION OF SCIENTIFIC FINDINGS: VISIONS AND CONCEPTS

The interrelationship of science and research on the one hand and growth and development on the other is an established reality. Research denotes the discovery of new facts and their utilization for problem solving. Failing to use scientific findings is equal to wasting human and material resources and to casting negative judgments toward the profitability of investments in science.

The development of information networks at every decision-making institution has resulted in easy access to scientific findings. One can have access to the most complex information in the shortest possible time.

In recent years, research and science production have been the focus of attention at many organizations. Government policies are based on strengthening and developing a productive research atmosphere. Developing countries, however, are at the beginning of this path. Because of a variety of factors, including software and hardware deficiencies, poor context and attitude, and inadequate human resources, the state of science and scientific applications in promoting the development of society is not yet at a desirable level. So far, much has been written about the relationship between decision-making, research, and information. Some reports concentrate on policy makers' and decision makers' inadequate attention to scientific findings. Others propose some ways for the enhancement of the impact of research on decision-making.

No studies have addressed the role of science utilization. Although international agencies have set some indicators for science production processes (e.g., development factors related to universities and research institutes), none has set criteria for science utilization.

An important point is the attitude and expectation of managers toward the term *utilization of scientific findings*. Some findings have closer affinity to the term *utilization* and are actually application oriented. Some are far from application and provide a basis for subsequent research that may take an applied direction.

Recognition of the reasons for failing to apply scientific findings will certainly play a part in increasing decision makers' and planners' attention to the application of such findings. Changing our attitudes toward the concept of utilization of scientific findings will revolutionize the role of research and scientific findings in the process of development. However, if we limit utilization exclusively to research and applied findings, then basic research, aimed at broadening the horizon of knowledge, will be put aside from the domain of knowledge. The existing view on the utilization of science and research limits utilization to a small window of time and place. There are many cases of scientific and research findings having profound, widespread influence on the views of the public at large and decision makers in particular, resulting in a significant role in public enlightenment.

Without well-defined criteria, as mentioned before, there cannot be any accurate assessment and reasonable judgment about the status of science utilization, its improvement, and its position in a society compared with societies in other countries.

In addition to the existing limited view of the utilization of scientific findings, there are numerous obstacles related to different areas which inhibit the desirable use of scientific findings. We may summarize some of these limitations, ranging from the irrelevancy of research topics to societal needs, poor quality of research, lack of relationship between scientific findings and societal demand, lack of belief in the research findings of decision makers and managers, and limitations in the acquisition of information.

TRENDS IN USING SCIENTIFIC FINDINGS IN DEVELOPMENT

The utilization of scientific findings denotes the payoff from scientific and research activities. Contrary to popular belief, utilization is quite a complex process and is the subject of dispute among researchers and decision makers. Today, the main debate centers around how to utilize scientific results better and more safely and thereby enhance the impact of scientific findings in the development of societies.

The production of science is a dynamic and lasting process with no definitive end. The publication of a scientific report may be considered the concluding point for a scientific finding, but the research process continues. Today's research is founded on yesterday's, and tomorrow's research relies on existing research results. While research is a continuum of related items, scientific findings are set forth in a broad spectrum of micro- and macro-applications. The micro-applications are at one end of the spectrum and comprise impacts on small societies, which may in some cases be extended to other, larger societies. The macro-applications are at the other end and embrace effects on decisions and policies at the national and, occasionally, international levels.

CONCLUSION

Studies of the utilization of scientific findings and the identification of constraints for their fulfillment at an optimum level for the acceleration of development will to some degree clarify how to approach a desirable level of utilization. The strategies for effective use of scientific findings necessitate the removal of already-known obstacles and limitations and, partially, the use of new methods for expediting this process. We may consider the process of science production and application as a unique system, a group of pieces and parts that are connected to each other and work together. In this system, some of the proposed strategies for the improvement of utilization are internal and some are external. For example, giving priority to scientific fields and programs that are required for the development of society is obviously important. Conducting research to provide qualitative scientific findings and taking into account the applicability of such findings is critical. Disseminating the culture of believing in science to managers and preventing them from making decisions on the basis of their own taste are significant objectives. Of course, providing essential financial resources as well as appropriate legal and organizational support is essential for the improvement of science production and utilization.

Considering the process of science production and utilization as one system has advantages that should be examined. Utilization would be addressed as one of the integrated parts in the promotion of science and, subsequently, development.

The author believes that existing research obstacles should be removed. However, more important in the improvement of utilization of scientific findings is encouraging managers to base their decisions on research findings. In cases where there exists a variety of research in one area, managers have learned to add research conclusions to each others' information and make use of meta-analysis rather than putting conclusions next to one another and accepting non-scientific considerations. On the other hand, in the absence or inaccessibility of scientific findings in an area, they have learned to avoid hasty decisions and have developed a sense of responsibility for requesting research in that area.

Finally, the most important approach that should be adopted in this context is the inclusion of science utilization in development indicators. Some criteria and benchmarks should be developed for the utilization of scientific findings in different areas. The most appropriate bodies for developing such criteria are international scientific agencies. These criteria in different science areas along with their defined indicators could provide a comprehensive strategy for measuring, evaluating, comparing, and improving the utilization of scientific findings in different societies.

BIBLIOGRAPHY

Aghazadeh, A. 1996. Ways of accommodating the utilization of research findings. *Quarterly Journal of Education Research* 23:6–14.

International Bureau of Education and National Institute for Educational Research. 1995. *Final Report of the International Meeting on Educational Reform and Educational Research: New Challenges in Linking Research, Information, and Decision-making, Tokyo, September 4–14.* Geneva: International Bureau of Education.

Funk, S. G., E. M. Tornquist, and M. T. Champagne. 1995. Barriers and facilitators of research utilization: An integrative review. *Nursing Clinics of North America* 30:395–407.

Ordonez, V., and Maclean, R. 1997. The impact of educational research on decision-making in education. In *Teachers, Teacher Education and Development: Report on an APEID Regional Meeting of Directors of Educational Research and Development Institutes in the Asia and the Pacific Region: Final Report of a Regional Meeting 7–15 July 1997, Kokuritsu Kyoiku Kenkyujo.* Tokyo: National Institute for Educational Research.

Saki, R. 2001. Utilization of research: A wide and multi-level concept. *Education Research Newsletter* (10):18–19.

Stone, D., S. Maxwell, and M. Keating. 2001. Bridging research and policy. Presented at an International Workshop Funded by the UK Department for International Development, Warwick University, Coventry, UK, July 16–17. Available at: http://www.gdnet.org/pdf/Bridging.pdf. Accessed August 21, 2007.

E

Ethics in Engineering as a Prerequisite for Technological Development of Societies

MEHDI BAHADORI and MAHMOOD YAGHOUBI
Academy of Sciences of Iran

Throughout history, engineers have been able to create a variety of technologies and solve diverse problems related to health and hygiene, treatment and cure of diseases, education, agriculture, housing, transportation, and other topics. This has been accomplished by engineers who avail themselves of the discoveries and scientific achievements of others and who use their own innovations and creativity. The outcome has been the provision of greater comfort and amenities for societies.

Along with these activities, engineers have also been able to create countless deadly weapons to support warmongers who may be able to annihilate or injure a large number of people in a short time.

Such activities of engineers have gone hand-in-hand with the pollution and destruction of the environment and the wasting of natural resources.

Engineers can, by using their creativity and innovation, solve problems and be the harbingers of many facilities and amenities for themselves and others. Having a strong sense of engineering ethics and morality, they can control their own activities, thus safeguarding the interests of societies and ensuring the health of the environment. Imbued with a sense of human values and engineering ethics, one can ensure peace of mind and inner satisfaction, ultimately creating a greater amount of personal happiness, the final goal of all human endeavors.

In this paper, a description of an engineer and the engineering profession is given. Then the role that science plays in this profession, the status of engineers in the development of society, and the importance that industrial countries have attached to engineering ethics are elaborated.

AN ENGINEERING OATH

An engineering oath has been proposed in Iran. Engineers can, at the time of graduation, sign the text of the engineering oath, which is as follows:

Profoundly conscious of the importance of the engineering profession in affecting the peace and welfare of human beings throughout the world, in protecting and safeguarding the environment against the hazards of pollution, in supporting my own sustainable joy and happiness, and by committing myself personally to this profession, I as an engineer hereby declare my sincere willingness to observe the following principles:

1. In all my engineering activities I shall observe the principles of honesty, precision, regularity, justice, speed in action, the interests of society, and the rights of my colleagues, and I shall be patient in reaching my objectives.
2. In all my engineering activities I shall observe the principles of health, safety, and the future of human beings, and I shall be kind, loving, and committed toward them.
3. In all my engineering activities I shall have self-confidence, curiosity, and perseverance, and I shall use creativity and innovation in solving problems assigned to me.
4. Concerning the duties assigned to me, I shall prove myself committed, conscientious, and restrained about my employer's secrets and collaborations.
5. In my engineering activities I shall economize in using water, energy, money, materials, equipment, time, and other national resources, and I shall have a keen sensitivity toward their use and avoid undue wastage.
6. In my engineering activities I shall try to limit harm to the environment as much as possible.
7. I shall endeavor to bring my engineering knowledge up to date and make myself conversant with the latest scientific and technical innovations and achievements, especially in such fields as hazards and safety, to economize in relation to materials, machinery, and systems being used, and to be fully knowledgeable in my designs.
8. In my engineering activities I shall try to create a working environment full of love and kindness. I shall endeavor to selflessly serve my fellow countrymen and look upon my colleagues as my dear brothers and sisters and love them all. I shall also try to cultivate human values both within myself as well as within others.
9. In my engineering activities I shall prove myself humble. I shall view the successes and breakthroughs achieved not solely as the result of my own initiatives but also as the result of the sincere cooperation of my colleagues. Hence, I shall always feel grateful toward them.
10. I shall impart my knowledge and experience to others in a spirit of loving kindness and selfless service.
11. In engineering designs I shall observe all standards, and I shall give highest priority to safety rules and the safety of society.

12. In my engineering activities I shall preserve a receptive mood and an open mind and sincerely accept the honest criticisms that my colleagues offer to me so as to overcome my shortcomings and mistakes, thus displaying my keen sense of appreciation for group or team work.
13. I shall refrain from any kind of malicious intention that may harm the prestige, honor, occupation, and material prosperity of my colleagues.
14. In my engineering activities I shall avoid taking any bribe. Similarly, I shall try to do away with other vices. In accepting rewards for my services, I shall charge payments that shall not exceed the bounds of decency and propriety.

ENGINEER AND ENGINEERING

Engineers, while making use of their creativity and innovation, can of course help solve problems. However, can they, by observing certain principles and ethical rules, effectively contribute toward overcoming and eradicating the pain and suffering that afflict humanity? Can they rid the environment of the pollution from which it suffers? Can they provide greater welfare for mankind?

The question that commands our attention is the purpose that welfare, material prosperity, and participation serve in a cut-throat competition that is daily assuming more complex and harrowing dimensions. Are we seeking greater physical comforts? Are engineers supposed to enhance material progress for the sole purpose of enjoying more sensual pleasures? Should science, technology, and innovation, with their breathless speed that has enabled them to utilize everything, be allowed to mercilessly follow their own present path without paying due regard to wisdom, virtue, morality, and human values? Does this take us nearer to the goal of being happy?

In the West, the word *engineer* in its present connotation has been in fashion since A.D. 1300. It comes from the Latin word *ingenium*. It has been recorded in various forms such as *ingeyno, engyn, engynne, ingenio.* For 700 years, different characteristics have been enumerated and kept in mind for engineers. Chief among them is that an engineer is a person possessing an innate talent and inborn genius, a human being that creates and is skilled in discovery, design, and innovation.

In the past two centuries, and especially in the twentieth and the beginning of the twenty-first century, the conceptions of engineers' responsibilities and services, which were valid until then, have undergone changes. With the enlargement of engineering boundaries and the establishment of different specialties, the domain of engineering services and occupations has been greatly broadened. This, in turn, has led to heavier responsibilities.

Engineering denotes people's ability in choosing, designing, innovating, planning, guiding, and manufacturing new or improved products. This ability is reflected in food and other agricultural products; manufacturing, reconstruction, and maintenance of machinery, tools, and buildings; and responses to many other needs of societies. Engineers have fundamentally transformed nature. Their activities have been accomplished by using resources such as materials and energy effectively, but at the same time, they have seriously polluted the environment.

Bearing the above comments in mind, engineering is a combination of knowledge and creativity. This can be acquired through education, training, research, and experience. It relies on the intelligence and talents of engineers for further innovation.

Of course, in the last half-century certain priorities for the knowledge and ability of engineers have been posed and have undergone modifications in accordance with changing needs. For instance, today, coping with environmental problems has assumed a far greater significance in the training of engineers.

Status of Engineers in National Development

Scientific and industrial development has passed through three important stages. These stages began in the years heralding the advent of the Industrial Revolution and continued until the second half of the twentieth century. Computer technology was introduced in the second half of the twentieth century. A third revolution relates to information technology and communications that emerged in the past two decades. Another revolution is unfolding: nanotechnology. Those who have been able to turn knowledge into innovation and innovation into new tools and novel instruments are engineers.

Today the process of continual, rapid change forms the most dominant feature in people's lives and its most salient characteristics. Creativity constitutes the bedrock of transformations and has played an extremely important role, and it is a characteristic of engineers. Many have called this century the Age of Creativity. However, creativity denuded of a philosophy based upon moral virtues, spiritual insight, wisdom, and conscience fails to propel the world toward a noble end and destiny.

Ethics

Bearing in mind the many different engineering activities, the advanced industrial societies, especially the United States, have resolved to pay more attention to ethics in science in general and engineering in particular. By observing these ethical precepts, engineers may control their own activities more effectively. The degree of attention paid to this matter can be seen in the establishment of centers relating to ethics, computer sites and communication lines for consultation in matters pertaining to engineering ethics, the preparation of ethics documents in companies and institutes and the training of engineers in interpreting the documents, and the publication of articles and books written about ethics in science and engineering. Ethics in engineering has also received attention in

Iran, as is evidenced by the number of publications (in Farsi) on the topic in the country in recent years.

In Iran, the most promising students select engineering subjects. These students, after struggling for 8,000 hours, acquire engineering knowledge and seek employment. By earning lucrative salaries during a period of about 40 years, these engineers are able to provide better and more comfortable lives for themselves and for their families and dependents. In the process of educating engineers in Iran as well as in other countries, universities instruct the young students on how to achieve a good living and be more comfortable.

How should these young specialists spend their lives after leaving the university? What ethical precepts and principles should they observe while engaged in their engineering activities so that they may enjoy life and be happier? For an efficient execution of our engineering responsibilities, we stand in need of certain principles that we call engineering ethics.

From the ethical point of view, engineers must possess the following: technical ethics that concern technical and scientific decisions; professional ethics that relate to dealings with other engineers, managers, workers, and support staff; and social morality that pertains to patriotic and humanitarian commitments.

How and where can one acquire these moral virtues? Is it essential that an engineer who has mastered all aspects of his studies and profession observe engineering ethics? Such questions constitute the very essence of our discussion in this paper.

We believe that the observance of ethical rules by engineers in all of their engineering activities is essential. Such observance makes for greater material prosperity for them in the long run. Happiness, peace of mind, and tranquility throughout life are thereby greatly enhanced.

We can compare the subject matter of ethics in the engineering profession to driving and the traffic signs at intersections. In relation to driving, the meaning of each and every green, yellow, and red light must be determined beforehand. The driver must know how to respond when confronted with them. Meanwhile, certain guidelines must also exist which may inform the driver of the punishments and the rate of fines imposed in case he disobeys the rules. Nevertheless, in spite of the existence of these punishments, the driver is morally bound to observe them so as not to encroach upon the rights of others and harm them or himself.

The same case holds for engineering ethics. An engineer should be instructed in ethics concerning the areas in which he can be active. He must be made aware of the situations or fields in which he must refrain from acting and be conscious of the consequences that lie in store for him in case he indulges in forbidden activities. However, more important is the imperative that an engineer should possess a noble sense of engineering ethics and human values. He should exercise restraint over his thoughts and actions.

Rules pertaining to legal and illegal activities must be imparted to students in universities. The penalties that will ensue from illegal activities must be determined through the country's laws. Most important, human values must be fostered in students of engineering and the country's engineers. It is the priceless possession of these sublime values that can ensure not only material welfare and prosperity but, more important, can contribute to greater inner satisfaction and sustainable happiness.

SCIENCE, INNOVATION, DEVELOPMENT, AND
THE ENGINEERING PROFESSION

The twenty-first century is an era of rapid changes and transformations. Advanced countries, through scientific discoveries and novel innovations, continue along their path of development unhampered. As a society becomes more advanced economically, culturally, and socially, it is consequently more capable of providing social security and mental tranquility for its citizens. The lack of such developments causes a host of social problems, such as increases in the rates of poverty and crime.

Attaining peace of mind, happiness, and prosperity is a major goal in any society. The growth and development of science and technology is a means toward the attainment of this objective. Reaching this goal requires a strong national determination and a true longing on the part of society. It also requires a sense of self-confidence and hard work by the members of society who can attain this goal.

The development of science and technology and human and moral characteristics are directly related. Unfortunately, the development of science and technology, with all the advantages and benefits that it has brought for man, has had a negative effect on moral and human values in society. The age of technology has caused a type of mechanization of human life and human behavior. It has caused people to drift far away from virtues and the accepted traditional values of society. The main reason for this is the industrial development of the country without adequate attention to the values dominating that society.

Engineers are one of the most important groups that directly contribute to the advancement of science and technology in a country. This group must possess certain characteristics in order to play its pivotal role as a connecting link or bridge between society and the industrial sector. The general belief is that an engineer must possess broad information skills that transcend his technical and technological skills. A good engineer, over and above being skilled in analyzing theories and their practical applications, must possess an analytical mind in critical situations. He must possess the ability to cope with prevailing work conditions, managerial skills, and the capacity to learn and to teach in the long run. He must also possess virtuous moral qualities.

More attention should to be paid during the training of engineers to the longevity of manufactured products and the engineers' responsibility toward the final performance of the product. Increased awareness concerning sustainable development and environmental issues, increased health and safety, and increased skills in group or team work must also be cultivated.

Today, the manufacturing of products and innovative services with better technology for competition at regional and world markets has become a necessity. In addition to drawing from the experiences of others to compete in various markets (especially in regional and world markets), creativity is an individual ability that can lead to a discovery or an innovative idea. Innovation is a complex and complicated process that renders a discovery or an idea into a marketable product or service. It is necessary to produce unique products and offer unparalleled services. This requires innovation and creativity.

A COURSE ON ETHICS FOR STUDENTS OF ENGINEERING

Along with teaching currently existing courses such as basic and engineering sciences in engineering colleges, it is important to instruct students in engineering ethics, too. Without such instruction, all that we succeed in accomplishing is to jam the minds of extremely intelligent young people with knowledge that is quickly forgotten. If, together with teaching engineering subjects, we also instruct students in engineering ethics, we will help to build a well-integrated character.

Most important in teaching engineering ethics and fostering the growth of human values is for the teachers themselves to possess such values. They should serve as models for the students who spend the best time of their lives in universities.

The first courses concerning engineering ethics were conducted in the United States in the 1960s. Initially, engineering ethics took the form of case studies of actions and decisions taken, either individually or collectively in relation to the engineering profession.

It is only appropriate that engineering departments in Iran make the necessary decision to offer such courses. This has to be led by a professor experienced in engineering. He should possess a blameless character and lofty human values. He should have close connections with industry on the one hand and technical innovation on a world scale on the other. The main subjects might include the following:

- history of engineering in the world,
- history of scientific and industrial revolutions in the world,
- working relations and industrial laws,
- economic and production relations,
- standards of design and productivity,
- professional ethics,
- human values and ethical engineering,
- environmental protection and sustainable development,

- globalization and the status of engineers in growth cycles, and
- relation of industry and university.

Each student should present a seminar on any of the topics listed above.

Students trained not only in engineering subjects but also in ethics can lead the way to developing a profession that responds to societal interests in a rapidly changing world.

F

Science and Society

REZA DAVARI ARDAKANI
Academy of Sciences of Iran

Modern science developed with the discoveries of Galileo, Copernicus, and Descartes. Usually, we have an abstract view of the advancement of such science up to the present era. There are rare cases when we consider the relationship of the scientist to historical conditions. It is a commonly held belief that science has reached its perfect stage and that everybody can learn, conduct, research, and benefit from science in the same way and at the same level. This view is particularly popular since it has an ethical element. That is, science is a value, and everyone can and should benefit from it. Such a view can solve no problem. It is not wrong, but it is superficial. The creation of science has always called for the existence of certain conditions and possibilities. Kant, as a teacher of enlightenment and liberalism, proclaimed the advent of reason, which could create a new science and related politics. Kant was well aware of what he was stating. He may have had the future in mind. He knew and taught others that each scientific discovery may not develop, last, and produce the expected benefits everywhere.

Science is a foundation that is closely linked to people's lives, relations, views, and behaviors. While growing, the tree of new science has challenged the earth and water and the spiritual and mental atmosphere of the Renaissance. It has gradually gained its place as the coordinator of conditions and the stabilizer of life order.

Science is said to have a method. Can any individual at any place reach the expected result of his/her research provided that he or she uses the right method? The conditions under which science is created and its benefits and dangers are the questions that philosophers, scientists, and politicians should address. This statement is true to some extent.

Even contemporary philosophers and scholars such as Gadamer, Foucault, Kuhn, Feyerabend, and Rorty neither take *method* seriously nor state that a scientist should ignore the rules. Rather they claim that science is not created by method. Here, however, we are not going to discuss the philosophy of science and criticize the views of philosophers.

The names of some contemporary scholars illustrate that the most prominent philosophers do not undermine science and its development. They warn us that if we have been trained in research methodology at school, we should not think that we will not become scientists. If we become scientists, we will not be able to lay the foundation of science without the collaboration of other scientists.

Everywhere in the world, scientists can establish scientific societies and cooperate in scientific research. Yet, the scientific society and its paradigms are not created by setting rules and taking administrative and official measures. Scientists will be successful in their scientific activities through teamwork and through conducting research on issues that are linked together. The scientific system is not a dispersed collection of research. Rather, it is a coordinated effort to resolve problems set forth by a scientific society. The American philosopher Richard Rorty rightly considers science a kind of solidarity that involves not only scientific subjects, problems, and scientists. Society and the life system are interrelated as well. This is particularly true in developed countries.

In the era of Galileo, Descartes, and even Newton, it was not likely that anyone considered science a solidarity. The growth of science and technology in a new society is not always constant and coordinated in all dimensions, although it has an affinity with the expansion of politics and law. Rorty, who considers science a solidarity, looks at science in the United States but apparently does not take seriously the 300- to 400-year-old history of challenge in Europe for the realization of such solidarity.

Obviously, the new science and society, in its essence and aptitude, have had solidarity since the very beginning. The history of science favors the coordination and solidarity with the history of society since the sixteenth century. In *New Atlantis*, Francis Bacon devised a society in which the rulers are scientists and the island is ruled by scientific rules. The design of the society is an anecdote and is the story of a solidarity dream for science and society.

Yet, when Bacon created his utopia, scientists did not rule society in actuality. The type of administration he described has not come into existence yet. This does not mean that Bacon's project has been put aside. The project was adjusted by his successors. Politics could not be formulated like physics, but Bacon's project was an introduction to the creation of a type of politics that would be coordinated with Galileo's and Newton's physics.

The development of science in the seventeenth and eighteenth centuries and the advent of the idea of a free and well-to-do society, without fear, enmity and war, is one dimension of the views of Thomas More, Francis Bacon's *New Atlantis*, Galileo, and Descartes' mathematical world. In the eighteenth century, the dominant idea was that a new sense of reason had come into existence that could lead science, politics, and society to freedom, peace, justice, and welfare. Now one can see that such a sense of reason as coordinating and leading is more or less problematic.

Rorty sees this affliction and shortcoming and hence does not want to call the eighteenth century the age of the founding of the history of science. Rorty is right, particularly in the sense that applying the scientific method was a necessary condition for finding the truth in the eighteenth century in the sense that the development of reason and science would result in the creation of a unique world. Science would be the pivotal point of agreement or disagreement among the people of the world. Now, science and technology, particularly information technology, are the factors that account for the uniformity of countries and regions, yet science has not developed or been equally useful at each and every place.

In developing countries, research faces great obstacles. These countries do not have sufficient budgets. The money that is allocated for research is not used in the right place and in an effective way because of the low level of technology in these countries. Even if they have a successful development plan, they still buy scientific and technological information. Consequently, in such countries research is mostly a tradition or a noble and formal profession. That is to say, in the developing world scientists are unable to conduct major and important research.

A glance at the list of articles published annually shows that the number of names belonging to the developing world is significant. Most likely, many names belong to scientists who have migrated from their homelands and now reside in locations where there are better conditions for conducting research. Nevertheless, they belong to developing nations and have an effective role in the development of the world's technology. Wherever the system of society is dominated by technology, scientists find their places easily. Since it is clear what types of research are given priority, the research conducted is that which is responsive to technology, economy, and politics. Here science is solidarity, and Rorty rightly does not consider science with the criterion of objectivity.

When objectivity is the absolute criterion—and of course one cannot easily ignore objectivity—all research is the same, and science is science wherever it is. This view is, of course, true to some degree. All research should be conducted on the basis of scientific criteria and consistent rules throughout the world, but the conditions for carrying out research vary from one place to another. We are used to limiting these conditions to available budgets, access to research facilities, and the absence of cultural and political obstacles. We scarcely recall that scientific movements would not have existed without the allocation of funds or political measures.

Europeans themselves began conducting research at a time when there was no political freedom (compared to the current freedom in Europe) nor was there any funding in their governments' budgets for research. Business and industry organizations did not take research seriously either. Obviously, under the present circumstances, states and governments as well as private institutions have no choice but to support technical-scientific research. These are necessary but not sufficient conditions.

The major points to consider for conducting research are the recognition of priorities and the existence of a comprehensive plan. Each scientist has his or her likes and dislikes. Their research will be fruitful and will add to science if their efforts complement other research and open the way for future efforts. Here the problem is that usually governments and powerful financial, economic, and military organizations prepare scientific plans. This may limit scientists' freedom. Authorities may also expect scientists to conduct research that is in line with their demands and interests. Clearly, influential institutions and organizations can direct research to such an extent that the research findings are rarely the same as the ones that scientists have expected, whereas eminent scientists have established their status with their great findings.

Hence the plan already mentioned is not the one that should be developed by individuals or institutions. This is in fact a development plan and comprises the issues that will bring about recession and despair in society if they are not resolved. If they are resolved, however, they will lay a foundation for setting forth and resolving other issues, science will flourish, and science and society will be solidified.

Note that the scientific and research issues of developing countries should not be confused with the issues of Europe and the United States. A scholar may rightly claim that the United States limits planning for research and the freedom of scientists and research. Such a scholar may also state that planning will result in the deviation of science and research from its natural path and will disturb its system. However, when order and solidarity are nonscientific, how should scientists find out what the major problems of society are? When there is no link between universities and technical, industrial, and economic institutions, how should researchers know what problems their countries' technology system, economy, and administration have? How should they collaborate in planning and resolving the problems?

At the outset, universities based in undeveloped and developing countries were merely educational centers. It has been only in the past two or three decades that research has found a place at universities. Still, the extent of the efficiency of the research conducted in developing countries is unknown.

One of the dominant and famous dogmas among the people of the world is that technology is the application of scientific rules. That is the reason physics, chemistry, and biology are termed basic sciences, as if scientific research is conducted first, and then engineers and technicians refer to the books and articles of the researchers of these sciences to create and use technology. In a sense, it is true that certain technology would not exist without conducting basic research, yet what should science deal with if not the technique? The issues of science are those of the technique.

Obviously, a scientist should not think of the benefits resulting from his or her research, nor should he or she think of its application. However, the scientist's problem comes from the field of technology, and his or her research findings will be used in this same field as well. Technique and technology are not preceded by scientific research, since technology and science are interrelated and inseparable. One can find no advanced technology in places where there is no research. In places where technology is at a minimum or in its initial stage, scientific research does not take place, and there is no need for it. If it does take place, it is in cooperation with the world's science. This is the case for many scientists and researchers of developing countries who deal with the universal problems of science (not merely their own societies' issues).

Actually, in a large part of the present world, research and development and development planning are inseparable. There may be great scientists in these regions, yet they may not choose the most important issues for their research. In the nineteenth and the first half of the twentieth century, if in Europe and the United States there was no plan for guiding research, and if scientists and universities had based their research on their own tastes, modern society would have had a different form of development.

We have seen that research has been more or less coordinated and complementary. When people are obliged to create their own science and society on the basis of a model existing elsewhere, they should be familiar with the internal order of that model and explore the causes of coordination and equilibrium or the origin of crises. This is very hard to do. If one succeeds in doing so, how can people be prepared to live in accordance with that model? How can one create a coordinated balance and the appropriate education, research, industry, and administration? If people do not change, what should be done with all those new products, and who benefits from them?

The major conditions for the creation of society and a scientific system are to have a sense of need, to belong to science, and to be prepared for and benefit from science.

Western thinkers are not very fond of exploring the world of undeveloped and developing countries. Recently, they have divided societies into modern and traditional, thereby complicating the problem more than before. Dividing the people of the world into modern and pre-modern is wrong, unjustifiable, and even dangerous. Undoubtedly, people living in different parts of the world have different traditions, and they are more or less attached to them. Presently, there is no place in the world that would have traditional society and order. Currently, no part of the world is devoid of the impact of modernity. No society is without the need of science and technology. However, science has not been attached to each society or to the spirit of people in the same way. Under the present conditions, it would be dangerous to divide societies into modern and traditional.

Since countries must develop at any cost, we should eliminate some traditions in order to pave the way for development and modernity. However, when this approach becomes a political agenda, it may end up a disaster. Societies can be divided into neither traditional and modern nor modern and premodern. The difference between the developing and undeveloped worlds and the Western world is that in the latter, science, life, politics, and so forth are all part and parcel of one body and have one spirit that leads them. However, in the other world, all of these have come from an external source and have been set without having a spirit for uniting and coordinating them. Here science is not solidarity. That is, research is not interrelated, nor does the collection of dispersed research have anything to do with the societies' order and politics, production and economy, and so forth. Though such a science is not yet solidarity, it should move in that direction.

It is not only a new pragmatic U.S. philosopher who considers science a solidarity. It is the ideal of modern society. Now, if we consider this an absolute and realized solidarity whose realization has been granted, we are too optimistic. Science, technology, politics, economics, and education are partially interrelated in modern society. However, there has always been the danger that science, research, and technology would not follow the law of solidarity and lead to a destructive path.

In the history of science, it has been stated that when Thales was studying the stars on a well, he slipped and fell into that well. A girl saw him, came to help him, and took him out of the well. The well was actually his telescope and an instrument for his research but turned into something that could take his life. However, beauty saved his life. It has been stated in different ways that art saves us from the probable toughness and hostility of science and politics. The story of Thales is meaningful.

We are not going to talk about art as a means of support for man. The world of information technology has narrowed down the space of art so much so that one cannot ask for help from art in case of danger. The developed world should not undermine the possibility of great risks. Nevertheless, the developing world should think about the solidarity and coordination of science, life, politics, and society. Of course, neither the developed or developing worlds should rely so heavily on science and technology that they lose their faith in art and thought.